International Perspectives on Aging

Volume 23

Series Editors

Jason L. Powell, Department of Social and Political Science, University of Chester, Chester, UK
Sheying Chen, Department of Public Administration, Pace University, New York, NY, USA

The study of aging is continuing to increase rapidly across multiple disciplines. This wide-ranging series on International Perspectives on Aging provides readers with much-needed comprehensive texts and critical perspectives on the latest research, policy, and practical developments. Both aging and globalization have become a reality of our times, yet a systematic effort of a global magnitude to address aging is yet to be seen. The series bridges the gaps in the literature and provides cutting-edge debate on new and traditional areas of comparative aging, all from an international perspective. More specifically, this book series on International Perspectives on Aging puts the spotlight on international and comparative studies of aging.

More information about this series at http://www.springer.com/series/8818

Marvin Formosa
Editor

The University of the Third Age and Active Ageing

European and Asian-Pacific Perspectives

 Springer

Editor
Marvin Formosa
Department of Gerontology
and Dementia Studies
Faculty for Social Wellbeing
University of Malta
Msida, Malta

ISSN 2197-5841   ISSN 2197-585X (electronic)
International Perspectives on Aging
ISBN 978-3-030-21514-9   ISBN 978-3-030-21515-6 (eBook)
https://doi.org/10.1007/978-3-030-21515-6

This Springer imprint is published by the registered company Springer Nature Switzerland AG
The registered company address is: Gewerbestrasse 11, 6330 Cham, Switzerland

This book is dedicated to all educational gerontologists for subscribing to the dictum that learning is living and that living is learning.

Foreword: The Largest University in the World

In the late 1980s, while I was working on my Ph.D. dissertation on the scientific and political history of gerontology, I became aware that the concept of ageism, having just surfaced academically in the 1970s, was undergoing a transformation. Critics began targeting 'positive' as well as 'negative' ageist images and narratives that divided or discriminated against older groups. The well-meaning gerontological vocabulary of 'positive', 'active' and 'productive' ageing was being appropriated by health and retirement markets capitalising on 'active lifestyles' populated by prosperous, independent and resourceful older adults. While such portrayals replaced the dowdy and depressing ones of the past, they also conveyed the new age of ageing as one in which people seemed to grow older without ageing and realistic or dignified representations of older people challenged by dependency or disability were absent from view. Centuries of moral debates about the universal virtues of ageing had morphed into a very commercialised one about 'successful' ageing.

Yet, there was a new age of ageing upon us with unprecedented demographic ramifications. The lengthening of the longevity curve and benefits of pension policies and medical technologies in the Western world had recreated retirement as an enlivened stage of life rather than entry to its ending. The affluent consumerism of post-war economies shaped a growing 'boomer' generation born between 1946 and 1964, who would develop distinctive lifestyles and tastes that extended their sense of rebellious youthfulness. Social thinkers, as they caught up with the new or post-traditional ageing, looked for language to describe it, such as the 'post-modern life course' (Featherstone & Hepworth, 1991), or more recently, the 'midlife industrial complex' (Cohen, 2012) and 'encore adulthood' (Moen, 2016). But no term stuck as well as 'the Third Age', as it was labelled in Europe and the UK, to describe a 'fresh map of life' (Laslett, 1989) wedged between the second age of youth and the fourth age of old age. The Third Age was characterised as a creative and active period overturning outmoded conventions about disengaging or slowing down during the retirement years. As Peter Laslett (founder of the British U3A movement) wrote, 'blanket phrases, which include "the Elderly," "Senior Citizens," "The Retired," and so on, have ceased to be appropriate now that the vital necessity of recognizing differentiation during that lengthy phase of life has become apparent' (1995: 10).

After my Ph.D., I began a project to identify new sites of Third Age emergence in travel, business, housing, fashion and the arts. I was looking for cultural manifestations of the new ageing that were not restricted to the positive/negative or active/dependent binaries churning through popular culture. I had heard about Universities of the Third Age (U3A), but it was my father-in-law, Colin Stamp, whose descriptions of his classes about religion and ethics organised by the U3A in London, who inspired me to look closer at U3A-style lifelong learning. The first U3A or *Université du Troisième age* was founded in Toulouse, France in 1973, followed by the establishment of the International Association of Universities of the Third Age (AIUTA). The first British U3A, inspired by the French model, began at Cambridge in 1982. As I learned about U3As in the UK and elsewhere, I realised that collectively they were the largest university in the world, yet were hardly acknowledged in gerontological research nor noticed by most traditional universities catering to younger students.

At first glance, U3As seemed like an unequivocal success, not only as inventive and inexpensive adult education, but also as a forum for local social reform. Branches are self-governed, decisions are collectively made, classes held in local spaces or members' homes, and no prerequisite experience or educational qualifications are required, nor are credentials, certificates, or degrees offered. U3A courses are based on interest and the love of learning, without the need for admission tests nor performances scores. I visited London and attended Colin's classes, interviewed some key U3A leaders such as Peter Laslett and Eric Midwinter, and read through U3A archives and newsletters whose access was kindly offered to me by the U3A London office. Clearly, the U3A experience for many people was the conduit through which they became Third Agers. After teaching university classes of young people for decades, I was fascinated to see how higher education could be retooled for middle-aged and older people too. Indeed, the 'senior power' behind the U3A movement shared many of its characteristics with other groups in which I was interested that looked to community resources to supplement diminishing social supports.

At the same time, by the turn of the millennium, Laslett and other figures in the U3A movement became accused of advocating 'third ageist' lifestyle priorities over economic and political issues, and for glorifying the positivity of the Third Age at the expense of the Fourth Age (Biggs, 1997; Blaikie, 1999; Gilleard & Higgs, 2000). Haim Hazan's (1996: 33) ethnography elaborated the interpersonal discourses and rituals through which Cambridge U3A participants fashioned their identities as a kind of 'buffer zone' between middle age and old age. He claimed that members were encouraged to 'abandon conventional parameters of time, space, and meaning and to reconstruct their own' (ibid., : 51). For Hazan and other U3A critics, this experimentation also encouraged an anti-ageing and death-denying atmosphere. Once again, a very positive outcome of the new ageing seemed to embody its own contradictions, as the energies, boldness and radicalism of the Third Age seemed to sharpen the boundary, even as it pushed it further away, between it and the Fourth Age, imagined as a reverse mirrored lifeworld darkened by loss, passivity and dependency (Gilleard & Higgs, 2010).

In my own work, I wrote about how the U3A administration handled tensions such as central versus local politics, research versus non-research activities and commercial versus educational interests (Katz & Laliberte-Rudman, 2005). But obviously much more needed to be written about Universities of the Third Age in order to encapsulate their global breadth as an expression of the Third Age itself, in all of its prospects and contradictions. Hence, we are very fortunate that Marvin Formosa, who has devoted his career to researching and consulting about older adult learning and the worldwide foundations of Universities of the Third Age, decided to put together this timely and comprehensive volume, *The University of the Third Age and Active Ageing: European and Asian-Pacific Perspectives.* The book's 21 chapters all contribute to the deep well of histories, experiences, structures, accomplishments and problems of national U3As. But each chapter can also be read separately as an absorbing story about special communities that have embraced the principles and exuberance of lifelong education and seek to understand, acknowledge and support it.

Overall, this book is a wonderful forum for its contributing authors, ranging from across European and Asian-Pacific countries to exchange ideas, learn from each other and respond to the critics. It is a massive amount of material; however, Formosa's exemplary editorial work, introductory and concluding chapters give readers the contexts and pathways to navigate the globalised U3A world as an exciting one of discovery. The book also leaves us with big questions about the meaning of lifelong education: How should it be organised and by whom? What will be the effects of digital and online learning technologies? Where should political activism and advocacy be part of U3A mandates? How can diversity be encouraged? Should U3As continue to identity with a Third Age against a Fourth Age? What does a shift in perspective from richer to poorer countries mean for the U3A movement? What other benefits to older adults can U3As provide beyond learning? How can U3As negotiate tensions between traditional and post-traditional cultures of ageing? These questions and more are integrated into Formosa's insightful conclusions gathered into his idea of U3A renewal; that it is time for the U3A movement to catch up to the myriad of changes occurring amongst ageing populations today, as it had once done as an innovative response to them in the past, and that we move from 'lifelong learning' to 'longlife learning'.

Engaging with these and other critical issues should make readers appreciate and enjoy this book as much as I have, seeing it as a tapestry of extraordinary research that offers to guide the U3A movement as it soon enters its fiftieth year, after which time its membership will swell with people for whom Universities of the Third Age will have always existed as an opportunity to never stop learning.

Peterborough, Canada   Stephen Katz

References

Biggs, S. (1997). Choosing not to be old? Masks, bodies and identity management in later life. *Ageing and Society, 17*(5), 553–570.

Blaikie, A. (1999). *Ageing and popular culture.* Cambridge: Cambridge University Press.

Cohen, P. (2012). *In our prime: The invention of middle age.* New York: Scribner.

Featherstone, M., & Hepworth, M. (1991). The mask of aging and the postmodern lifecourse. In M. Featherstone, M. Hepworth, & B.S. Turner (Eds.), *The body: Social process and cultural theory* (pp. 371–398). Thousand Oaks, CA: Sage.

Gilleard, C., & Higgs, P. (2000). *Cultures of ageing: Self, citizen and the body.* London: Prentice Hall.

Gilleard, C., & Higgs, P. (2010). Theorizing the fourth age: Ageing without agency. *Aging and Mental Health, 14*(2), 121–128.

Hazan, H. (1996). *From first principles: An experiment in ageing.* Westport, CT: Bergin & Harvey.

Katz, S., & Laliberte-Rudman, D. (2005). Exemplars of retirement: Identity and agency between lifestyle and social movement. In S. Katz (Ed.), *Cultural aging: Life course, lifestyle and senior worlds* (pp. 140–160). Peterborough, ON: Broadview Press.

Laslett, P. (1989). *A fresh map of life: The emergence of the third age.* London: Weidenfeld and Nicolson.

Laslett, P. (1995). The third age and the disappearance of old age. In E. Heikkinen, J. Kuuisinen & I. Ruoppila (Eds.), *Preparation for aging* (pp. 9–16) New York: Plenum.

Moen, P. (2016). *Encore adulthood: Boomers on the edge of risk, renewal, & purpose.* Oxford: Oxford University Press.

Preface

My personal and academic interest in Universities of the Third Age (U3As) can be traced back to the mid-1990s when, as a postgraduate student at the University of Malta, I was researching the field of adult education, which led me to the field of educational gerontology. At that time, without Internet search engines at our disposal, rummaging for academic material on the field of older adult learning was truly a creative and resilient quest. I remember writing letters—the traditional way —to Frank Glendenning, Malcolm Johnson, Keith Percy, Alexandra Withnall, Chris Phillipson and other leading academics who I felt could provide me with guidance and advice. All responded to my letters, and some even sent me a range of grey literature, for which I am still very grateful, since at that time travel to the UK, and conference fees were beyond my budget. Encouraged by such a response, I carried out my Masters' dissertation in sociology on older adult learning, by conducting a case study of Maltese U3A. The dissertation was well-received and generated a number of articles in peer-reviewed journals, all of which endeavoured to inject a critical twist to the field of older adult learning and U3As in particular. Despite a short hiatus, as I channelled all my energies into completing a Ph.D. in gerontology, researching class dynamics in later life, my academic gaze never left this area, as I remained marvelled by the sheer expansion of the number of older persons engaged in formal, non-formal and informal avenues of learning. Indeed, since the millennium, the opportunities for lifelong learning in later life have proliferated in an unprecedented manner, notwithstanding that many governments have cut back on their public spending on adult education, and that state budgets on education rarely take into consideration the learning needs and interests of older citizens.

This book is, of course, downright testimony that no other enterprise in older adult learning has been as successful as the U3A. Since its inception in 1973, the U3A defied the odds, to now include millions of learners in its fold, and to be present across all inhabited continents, to the extent that today one speaks of a 'U3A movement'. Although comparative statistics are lacking, Australia boasted some 300 U3As, with a membership of around the 100,000 mark, whilst its neighbour New Zealand held 84 U3As with members within the 25 Auckland U3A community numbering 3,719 in 2017. Figures for Britain reached over 1000 U3As and over 400,000 members in 2018, and a 2013 Interest Group Survey revealed that there are in excess of 36,000 U3A interest groups in the region. In the Asian continent, China alone owned 60,867 U3As and around 7,643,100 members in 2015. The genesis of this book emerged from an intention to take stock of such a state of circumstances—namely to detail the movement's origins, development and contemporary equilibrium; chart the movement's impact on older learners' levels of physical, emotional and social well-being; as well as evaluate the movement's track record in bringing improved levels of empowerment in later life. Whilst the initial objective was to produce a more comprehensive, international, compendium of chapters, the choice of limiting this book to solely the European and Asian-Pacific settings was based on the fact that whilst U3As in these two continental locations are both highly popular and firmly entrenched in the social fabric, there are only a few U3As in Africa and they are completely absent in the USA. At the same time, language barriers made it difficult for me to locate apposite collaborators from South America. One hopes that such limitations can be rectified in future publications in the foreseeable future.

The choice of the term 'active ageing' in the title of the publication was not a coincidental one. The concept of 'lifelong learning' is presently positioned as one of the four key pillars underlying the discourse of 'active ageing', so that nowadays these two notions have become increasingly interfaced, overlapping and pervasive, to the extent that both the United Nations and the World Health Organization recognise that neither active ageing nor lifelong learning is possible at the expense of the other. Acknowledging that lifelong learning and active ageing resemble the two different sides of the same coin—in its policy document *It is never too late to learn* (European Commission, 2006)—the European Union also drew attention to the need for 'active ageing' policies addressing the need for learning both before and after retiring from formal work. Indeed, research has long confirmed that continued learning is a key vehicle for active ageing (Boulton-Lewis & Buys, 2015; Formosa, 2016), whereby it can enable older persons

> …gain socio-economic, psychological, and socio-political resources, all of which in turn lead to a healthier life. In addition to better health, older adults engaged in lifelong learning are found to have positive experiences in at least one of the following areas - enjoyment of life, confidence, self-concept, self-satisfaction, and the ability to cope…learning keeps older people involved in enjoying and living life fully.
> Tam, 2012 : 165

At the same time, there is some research which provides evidence to support that engagement in learning slows mental decline, as learning boosts brain activity which has the potential to improve neuroplasticity, neural development and new brain cells (Valenzuela, 2009).

A large number of people have played a vital role during the writing of the book. Much gratitude goes to the authors of the chapters in this book. There is no doubt that the success of this volume is due primarily to the commitment with which the chapter authors accepted their obligations and the goodwill with which they responded to editorial criticism and suggestions. Amongst the chapters' authors, special gratitude goes to Cameron Richards for advising me, correctly with hindsight, to include a 'Pacific' dimension in the book as primarily the focus was an entirely 'European-Asian' one. The book has also benefitted from the input of many colleagues and friends whom I have met during many conferences and study visits throughout the past two decades, especially throughout the annual conference organised by the European Society for Research on the Education of Adults (ESREA), on Education and Learning of Older Adults (ELOA). During such conferences, my debates with Bernhard Schmidt-Hertha, António Fragoso, and Sabina Jelenc Krašovec always served as sources of fresh knowledge, and to recognise various facets of older adult learning through new perspectives. Gratitude also goes to Katie Chabalko and the production team at Springer Publishing for being ever present and available throughout this book's production process.

Msida, Malta   Marvin Formosa

References

Boulton-Lewis, G. M., & Buys, L. L. (2015). Learning choices, older Australians and active ageing. *Educational Gerontology, 41*(11), 757–766.

European Commission. (2006). *Adult learning: It is never too late to learn.* Brussels, Belgium: European Commission.

Formosa, M. (2016). Malta. In B. Findsen & M. Formosa (Eds.), *International perspectives on older adult education: Research, policies, practices* (pp. 261–272). Dordrecht, Netherlands: Springer.

Tam, M. (2012). Active ageing, active learning: Elder learning in Hong Kong. In G. Boulton-Lewis, & M. Tam (Eds.), *Active ageing, active learning: Issues and challenges* (pp. 163–174). Dordrecht, Netherlands: Springer.

Valenzuela, M. J. (2009). *It's never too late to change your mind.* Sydney: ABC Bo.

Contents

Editor and Contributors

About the Editor

Marvin Formosa is Associate Professor of Gerontology at the University of Malta where he is Head of the Department of Gerontology and Dementia Studies, Faculty for Social Wellbeing, and contributes to teaching on active ageing, transformative ageing policy, and educational gerontology. He holds the posts of Chairperson of the National Commission for Active Ageing (Malta), Rector's Delegate for the University of the Third Age (Malta), and Director of the International Institute on Ageing, United Nations, Malta (INIA). He directed a number of international training programmes in gerontology, geriatrics and dementia care in the Philippines, China, India, Turkey, Malaysia, Belarus, Kenya, Argentina, Azerbaijan and the Russian Federation. He has published extensively across a range of interests, most notably on active ageing, critical gerontology, Universities of the Third Age and older adult learning. Recent publications included *International perspectives on older adult education* (with Brian Findsen, 2016), *Population ageing in Turkey* (with Yeşim Gökçe Kutsal, 2017), and *Active and healthy ageing: Gerontological and geriatric inquiries* (2018). He holds the posts of Editor-in-Chief of the International Journal on Ageing in Developing Countries, Country Team Leader (Malta) of the Survey of Health, Ageing, and Retirement in Europe (SHARE), and President of the Maltese Association of Gerontology and Geriatrics. e-mail: marvin.formosa@um.edu.mt

Contributors

Siti Aisyah Nor Akahbar Laboratory of Medical Gerontology, Malaysian Research Institute on Ageing, Universiti Putra Malaysia, Serdang, Malaysia

Barbara Baschiera University of Malta, Msida, Malta

Cecilia Bjursell School of Education and Communication, Jönköping University, Jonkoping, Sweden

Maya Abi Chahine American University of Beirut, Beirut, Lebanon

Sen Tyng Chai Laboratory of Social Gerontology, Malaysian Research Institute on Ageing, Universiti Putra Malaysia, Serdang, Malaysia

Ernest Chui Department of Social Work and Social Administration, Sau Po Center on Ageing, University of Hong Kong, Pokfulam, Hong Kong

Karen Evans UCL Institute of Education, London, UK

Brian Findsen Division of Education, University of Waikato, Hamilton, New Zealand

Marvin Formosa University of Malta, Msida, Malta

Hans Kristján Guðmundsson U3A, Reykjavík, Reykjavík, Iceland

Tengku Aizan Hamid Malaysian Research Institute on Ageing, Universiti Putra Malaysia, Serdang, Malaysia

Chin-Shan Huang National Chung Cheng University, Chiayi, Taiwan

Soo-Koung Jun Namseoul University, Cheonan, South Korea

Thomas Kuan U3A Singapore, Singapore, Singapore

Shu-Hsin Kuo Toko University, Puzi City, Taiwan

Ainslie Lamb U3A Network New South Wales, Sydney, Australia

Jolanta Maćkowicz Pedagogical University of Cracow, Kraków, Poland

Jittra Makaphol Silpakorn University, Bangkok, Thailand

Gulnara Minnigaleeva National State Research University Higher School of Economics, Moscow, Russia

Keith Percy Lancaster University, Lancaster, UK

Nur Aira Abd Rahim Department of Professional Development and Continuing Education, Faculty of Educational Studies, Universiti Putra Malaysia, Serdang, Malaysia

Cameron Richards Southern Cross University, Lismore, Australia

Bernhard Schmidt-Hertha University of Tuebingen, Tuebingen, Germany

Abla Mehio Sibai American University of Beirut, Beirut, Lebanon

Maureen Tam Education University of Hong Kong (EdUHK), Tai Po, Hong Kong

François Vellas University of Toulouse Capitole, Toulouse, France

Feliciano Villar University of Barcelona, Barcelona, Spain

Joanna Wnęk-Gozdek Pedagogical University of Cracow, Kraków, Poland

Noor Syamilah Zakaria Department of Counsellor Education and Counselling Psychology, Faculty of Educational Studies, Universiti Putra Malaysia, Serdang, Malaysia

Xinyi Zhao School of Health Humanities, Peking University, Beijing, China

Part I
The Background Context

Chapter 1
Active Ageing Through Lifelong Learning: The University of the Third Age

Marvin Formosa

From Active Ageing...

The notion that remaining active in later life holds various benefits for older persons has a long lineage in ageing studies. In the 1960s, activity theorists argued that ageing successfully depended on the maintenance of activity patterns and values typical of middle age (Havighurst, 1963). Although this standpoint achieved much acclaim initially, it was eventually criticised as idealistic and unrealistic. Many older adults do not only face biological limitations in keeping up with a middle-age lifestyle, but are also shackled by economic, political and social structures of society—most notably, mandatory retirement—that constrain and prevent them from remaining active (Walker, 1980, 1981). As a result, the 1980s experienced a shift in ageing studies away from activity rationales, and towards a focus on the political economy of old age and the distribution of resources in later life (Estes, 1979; Guillemard, 1981; Phillipson, 1982). However, this trend reversed itself during the 1990s as the World Health Organization (WHO) (1994) emphasised the advantageous link between activity and health. During the United Nations' International Year of Older Persons (1999), the WHO acclaimed a strong association between activity and opportunities for healthy living in later life and highlighted the need to create opportunities for older people to age actively (Kalache, 1999).

The WHO's discourse on 'active ageing' found immediate support in both policy and popular circles, as it surfaced at a time when the issue of population ageing was forcing the dismantling of the traditional conception of the life course that equated the oldest phase of life with frailty and liability (Boudiny & Mortelmans, 2011). As this standpoint focused on encouraging the participation of older adults in society, whilst emphasising the resources and knowledge that older people possess, many an advocacy group took on 'active ageing' as their *cri de guerre*. Heartened by such

M. Formosa (✉)
University of Malta, Msida, Malta
e-mail: marvin.formosa@um.edu.mt

© Springer Nature Switzerland AG 2019
M. Formosa (ed.), *The University of the Third Age and Active Ageing*, International Perspectives on Aging 23, https://doi.org/10.1007/978-3-030-21515-6_1

developments, the WHO presented a policy framework on active ageing at the Second
World Assembly on Ageing, wherein 'active ageing' was defined as

> ...the process of optimizing opportunities for health, participation and security in order to
> enhance quality of life as people age. Active ageing applies to both individuals and population
> groups...The word "active" refers to continuing participation in social, economic, cultural,
> spiritual and civic affairs, not just the ability to be physically active or to participate in the
> labour force.
>
> (WHO, 2002: 12)

The WHO's sponsorship of an 'active ageing' discourse was, in many ways, a
positive move in the right direction, transcending ageing policy from 'the placid
backwaters of politics into the mainstream of economic, social and cultural debate'
(Salter & Salter, 2018: 1069). For Walker, the WHO made two important contributions to global discourses on active ageing:

> First, the WHO policy added further weight to the case for a refocusing of active aging away
> from employment and toward a consideration of all of the different factors that contribute
> to well-being. Specifically, it argued for the linkage, in policy terms, between employment,
> health, and participation and echoed the similar case made within the European Union.
> Second, and again along similar lines as the contributions of the European Commission and
> European scientists, it emphasized the critical importance of a life-course perspective. In
> other words, to prevent some of the negative consequences associated with later life, it is
> essential to influence individual behavior and its policy context at earlier stages of the life
> course.
>
> (Walker, 2009: 84)

In the original policy framework, WHO (2002) hinged active ageing upon the
three pillars of health, participation and security. Primarily, when the risk factors
for chronic diseases and functional decline were kept low, people were expected to
enjoy 'both a longer quantity and quality of life; they will remain healthy and able
to manage their own lives as they grow older; fewer older adults will need costly
medical treatment and care services' (ibid.: 45–6). With regard to participation, the
WHO advocated that when employment, education, health and social services support older persons' participation in socio-economic, cultural and spiritual activities,
they 'make a productive contribution to society in both paid and unpaid activities
as they age' (ibid.: 46). Finally, it was contended that when programmes address
the social, financial and physical security needs of older people, they are 'ensured
of protection, dignity and care in the event that they are no longer able to support
and protect themselves' (ibid.). In due course, and highly noteworthy considering
the ethos of this edited volume, an updated report of the WHO's landmark document
added lifelong learning as the fourth pillar of active ageing, The notion of 'active
ageing' was thus redefined as the 'process of optimizing opportunities for health,
lifelong learning, participation and security in order to enhance the quality of life as
people age' (International Longevity Centre Brazil, 2015: 43—italics added). Whilst
in 2002 the active ageing framework did acknowledge that lifelong learning—along
with formal education and literacy—is an important factor that facilitates participation, health and security as people grow older, in 2015 the placing of lifelong

learning as a fourth pillar of the official 'active ageing' discourse signalled strongly that learning is important not only to productive ageing, but also to the reinforcement of well-being in later life. To cite the updated report, lifelong learning

> …is a pillar that supports all other pillars of Active Ageing. It equips us to stay healthy, and remain relevant and engaged in society. It therefore empowers and gives greater assurance to personal security. At the societal level, people in all walks of life and at all ages who are informed and in possession of current skills contribute to economic competitiveness, employment, sustainable social protection and citizen participation.
>
> (International Longevity Centre Brazil, 2015: 48)

The integration of 'lifelong learning' into active ageing discourse functioned to safeguard the right of persons to age positively since the key role that learning may play in promoting quality of life and well-being in later life has long been recognised by academics and policy makers (Findsen & Formosa, 2016a). As a recent study concluded,

> …older adults' participation [in learning] is independently and positively associated with their psychological wellbeing, even among those typically classified as 'vulnerable'… continuous participation in non-formal lifelong learning may help sustain older adults' psychological wellbeing. It provides older learners, even those who are most vulnerable, with a compensatory strategy to strengthen their reserve capacities, allowing them to be autonomous and fulfilled…
>
> (Narushima, Liu, & Diestelkamp, 2018: 651)

It is also noteworthy that the results of this research investigation highlighted the significance of strategic and unequivocal promotion of community-based non-formal lifelong learning opportunities, for the encouragement of inclusive, equitable and caring active ageing societies. This is, of course, not the same as saying that the WHO's discourse on active ageing is devoid of lacunae and limitations, and despite its popular status in ageing policy, the 'active ageing' paradigm also received its fair share of criticism. Boudiny (2013) argued that the active ageing framework tends to be used in a narrow manner, to promote physical activity and prolong labour participation, and often used interchangeably with 'productive ageing' and 'healthy ageing'. Moreover, the active ageing paradigm has been found to be incomplete as far as frail and vulnerable persons are concerned, especially with respect to long-term care contexts (van Malderen, Mets, De Vriendt, & Gorus, 2013). On similar lines, Paz, Doron, and Tur-Sinai (2018) contended that a gender-based approach is missing within the existing active ageing framework. Nevertheless, despite such possible lacunae there can be no doubt as to the contribution of the active ageing discourse towards the challenging of stereotypes which characterise later life as a period of passivity and dependency, by placing an opposing emphasis on autonomy and participation (Formosa, 2017).

...to Lifelong Learning in Later Life

Lifelong learning in later life refers to the process in which older adults, 'individually and in association with others, engage in direct encounter and then purposefully reflect upon, validate, transform, give personal meaning to and seek to integrate their ways of knowing' (Mercken, 2010: 9). Although statistical research on the participation rates of older persons in learning activities remains sparse, a critical review of available literature elicited three constant findings—namely a lower percentage of elder learners compared to younger peers, a sharp decline of participation as people reached their seventieth decade, and that typical learners are middle-class women (Findsen & Formosa, 2016b). In the USA, the 2005 National Household Education Survey found that 23% of adults aged 65-plus participated in a non-accredited learning activity organised by community or business institutions in the previous year (O'Donnell, 2006). In Europe, Eurostat (2018) reported that in 2017, across the EU-28 Member States, 4.9% of the population aged 55–74 participated in formal and non-formal educational activities, although this figure reaches 5.8% if one only takes into consideration the 15 EU-Member States. Research on the participation of older adults in solely non-formal learning activities elicits higher results, as in the case of the survey conducted by the Spanish National Institute of Statistics which found that 22.8% of the sample aged 55–74 participated in a learning activity in the previous 12 months (Villar & Celdrán, 2013). With regard to the preferred subject matter of older learners, older adults are generally stereotyped as favouring 'expressive' over 'instrumental' forms of learning. However, whilst the interest of older adults to engage in expressive learning programmes can never be overstated, many older learners are highly interested to learn computer skills and subjects in the natural and physical sciences such as biology, marine habitats, geology and astronomy (Talmage et al., 2015).

A key debate in older adult learning is concerned not with 'whether we can or cannot teach or retrain an older adult' but 'to what end?' and 'why?. Primarily, late-life learning was commended for assisting adults to adjust to transformations that accompany 'old age' such as decreasing physical strength and health, the retirement transition, reduced income, death of spouse, and changing social and civic obligations (McClusky, 1974). Glendenning and Battersby (1990), and subsequently Formosa (2011, 2018), posited a more radical agenda and bestowed late-life learning the task of achieving the 'liberation of elders'—that is, empowering older persons with the advocacy skills necessary to counteract the social and financial disadvantages brought on by neoliberal politics of ageing. From a humanistic point of view, learning in later life was perceived as a 'personal quest', a necessary activity if older adults are to achieve the potential within them (Percy, 1990). This rationale prioritised 'process' over 'content' by stressing that the role of an educator "is to facilitate the process of learning for the learner" rather than "persuade him [sic] to social action or to be dissatisfied if a certain political awareness is not achieved" (ibid.: 236). Finally, transcendence rationales argued that learning must not let adults forget that they are 'old' and should enable learners 'to know themselves as a whole, as they really are, in the light of finitude and at the horizon of death' (Moody, 1990: 37). Consequently, lifelong learning in later life arises as an opportunity to explore goals that younger peers are too busy to pursue, such as developing a reflective mode of thinking and contemplating the meaning of life. Whilst these stances hinge the rationale for lifelong learning in later life on different tangents, each has positive contributions to make, and if taken together there will be no doubt of the potential of older adult learning to bring the 'need for economic progress and social inclusiveness in tandem with recognition of individual desires for personal development and growth as people age' (Withnall, 2010: 116). Indeed, the blending of active ageing and lifelong learning discourses leads to the recognition that learning takes place in a variety of contexts—such as formal classrooms, self-directed learning, voluntary organisations, residential and nursing homes and intergenerational settings—to the possible outcome of three key types of benefits:

- *Psychological well-being.* The midlife-to-ageing transition brings a number of possible changes such as decreasing physical strength and health, exit from the labour market, reduced income, widowhood and changing social and civic obligations, amongst others (Maginess, 2017). Late-life learning has the potential to mitigate against such turning points and enable older persons to achieve higher levels of life enjoyment, self-confidence, self-esteem, self-satisfaction and wide-ranging coping strategies (von Humboldt, 2016).
- *Social inclusion.* Ageing has an invariable effect on all incumbents in increasing their risk of social exclusion on the basis on their chronological age, and especially more so for vulnerable members on the basis of their gender, social class, geographical location, sexuality and ethnic status (Hafford-Letchfield, 2016). Older adult learning has the potential to adopt a wide-participation agenda and bring about improved levels of social support, social networking and social solidarity amongst older persons (Findsen & Formosa, 2011).
- *Empowerment.* Lifelong learning in later life can arise as an expression of older persons' need to regain power over their own lives (Glendenning & Battersby, 1990), meaning that this type of learning empowers learners through exposure to the politics of daily living (Hachem, Nikkola, & Zaidan, 2017). Indeed, education in later life should 'illuminate the social and political rights of older people' (Hafford-Letchfield, 2014: 438), to equip older adults to claim their social rights and induce social change (Formosa, 2012).

Indeed, many a research article has documented how opportunities for lifelong learning in later life act as a catalyst for healthier and socially engaged lives in old age, as well as improving older persons' opportunities to either remain in or re-enter the labour market (Findsen, Mcewan, & Mccullough, 2011; Harris & Ramos, 2013). In brief, older adult learning was found to play a vital role in maintaining cognitive functioning and capability (Boulton-Lewis & Tam, 2012; Jenkins & Mostafa, 2015); improving individuals' social relations as they interact with, and learn from, same-aged or older/younger peers (Åberg, 2016; Kimberley, Golding, & Simons, 2016); keeping up with computer and Internet-related developments (González-Palau et al., 2014; de Palo, Limone, Monacis, Ceglie, & Sinatra, 2018); supporting older adults in gaining socio-economic resources in retirement (Lido, Osborne, Livingston, Thakuriah, & Sila-Nowicka, 2016; Talmage, Mark, Slowey, & Knopf, 2016); and strengthening personal development by bolstering empowerment, self-esteem and confidence, and positive outlooks towards life (Narushima et al., 2018; Wang et al., 2018).

The University of the Third Age

Opportunities for lifelong learning in later life hold a rich tradition in many continents and countries. In the North American hemisphere, the inaugural lifelong learning institute targeting older persons, the Institute for Learning in Retirement, was established as early as 1962 and was followed by the establishment of Shepherds Centers, the Fromm Institute for Lifelong Learning, Osher Lifelong Learning Institutes, Harvard Institute for Learning in Retirement and the indefatigable Elderhostel (Road Scholar since 2011). In Europe, key organisations involved in providing opportunities for late-life learning include the University Programmes for Older People and the Universities of the Third Age (U3As) which is, of course, the focus of this collection of chapters.

The origins of U3As can be traced to legislation passed by the French government in 1968 which obliged universities to become responsible for the provision of lifelong education. In the summer of 1972, this legislation inspired Pierre Vellas to coordinate—at the University of Toulouse—a summer programme of lectures, guided tours and other cultural activities, for retirees (Radcliffe, 1984). When the programme came to an end, the enthusiasm of participants showed no signs of abating, so that Vellas (1997) planned a new series of learning programmes for the ensuing year under the name of the University of the Third Age. In Funnell's words,

> In February 1973 a proposal to create the University of the Third Age of Toulouse was put to the Administrative Council of the International Studies and Development Faculty which had representatives of the professors, the students and the administrators as well as three external members with important international responsibilities: the directors-general of the World Health Organisation (WHO), the International Labour Organisation (ILO) and UNESCO. The goal was to investigate what the university could do to improve the quality of life and state of health of the elderly [sic]. The programme was adopted unanimously, although without any specific budget…
>
> (Funnell, 2017: 120)

The first U3A was open to anyone who had reached statutory retirement age at that time, who was willing to pay a nominal fee. Learning activities were scheduled for daylight hours, five days a week, for eight or nine months of the year. The first curriculum at Toulouse focused on a range of gerontological subjects, although in subsequent years, subject content mainly focused on humanities. Although lectures were combined with debates, field trips and recreational opportunities, the French academic maxim of 'teachers lecture, students listen' was constantly upheld. The Toulouse initiative struck a rich vein of motivation so that, just three years later,

U3As were already established in Belgium, Switzerland, Poland, Italy, Spain and Quebec in Canada. Such popularity was due to the fact that the U3A movement, in marked contrast to the tradition of centralised educational management, provided an opportunity to sow the first seeds of educational innovation and reform. The first U3A in Britain was established in Cambridge, in July 1981, and quickly replicated in other cities and towns. The British version underwent a substantial change compared to the original French model and was founded upon three principles (Laslett, 1989): first, a Third Age principle where members are expected to take some responsibility, in particular, to ensure that any potential members might have the opportunity to join the U3A ranks; second, a self-help learning principle as members gather and immobilise themselves into learning circles, utilising as broad a selection of themes as they think fit; and finally, a mutual-aid principle as each U3A is organised on the co-operative precept, thus being totally autonomous and self-sufficient.

The U3A developed from modest circumstances in Toulouse, in France, to a global movement encompassing thousands of centres and millions of members (Formosa, 2014). U3As are characterised by a strong self-governing trait that renders efforts to arrive at a reliable and valid definition excruciating. However, U3As can be loosely defined as 'socio-cultural centres where senior citizens may acquire new knowledge of significant issues, or validate the knowledge which they already possess, in an agreeable milieu and in accordance with easy and acceptable methods, with the objective of preserving their vitality and participating in the life of the community' (Midwinter, 1984: 68). As its title postulates, the U3A's target audience are people in the Third Age phase of their life course. Whilst some centres put age 60 and over as a prerequisite for membership, others however adopt a more flexible approach by opening membership to all persons above the age of 50. The U3A movement has not only withstood the test of time but is also marked by an extensive increase of centres and members all over the globe. Although comparative statistics are lacking, Australia included some 300 U3As, with a membership of around the 100,000 mark (U3A Alliance Australia, 2017), whilst its neighbour New Zealand held 84 U3As with the members of the 25 Auckland U3A communities numbering 3719 in 2017 (Findsen, personal communication). Figures for Britain reached over 1000 U3As (400,000 members) in 2018 (U3A, 2018), and a 2013 Interest Group Survey revealed that there are in excess of 36,000 U3A interest groups in the region (Withnall, 2016). Whilst—as expected—walking, history and 'going out' have been listed as the most popular activities, bus restoration, Druidism and unsolved murder mysteries were other unusual subjects on offer (ibid.). In the Asian continent, China alone included 60,867 U3As and around 7,643,100 members in 2015 (Teaching Study Department Guangzhou Elderly University, 2017). At the same time, U3As are also no exception to the e-learning revolution. Initially, the scope of online courses was to reach out to older persons who could not join their peers in a classroom setting such as those living in remote areas and the homebound (Swindell, 2011). However, the coming of the Web 2.0 Internet revolution in union with steep increases in digital competence in later life made virtual U3As—such as U3A Online—increasingly popular with Third Agers (Formosa, 2014).

To dwell on the movement's diverse governing structures, goals and curricular objectives is highly pre-emptive, considering that the rest of the chapters in this book address the unique nuances of U3As in European and Asian-Pacific contexts. However, it is noteworthy to point out, even if briefly, the extensive research lauding the positive impacts of the U3A movement on older persons, and the wider community in general. Although one finds no longitudinal research on the relationship between U3A membership, on the one hand, and improvement in physical and cognitive well-being, on the other, various cross-sectional studies outlined how U3As bring about direct benefits to their members' overall welfare (Maniecka-Bryła, Gajewska, Burzynska, & Byrla, 2013; Niedzielska et al., 2017; Zielinska-Wieczkowska, Ciem-noczołowski, Kedziora-Kornatowska, & Muszali, 2013). An association between participation in U3As and improved levels of self-assurance, self-satisfaction, self-esteem and sense of coherence, on the one hand, and a decline of depressive and anxiety symptoms, on the other, is a frequent result in both survey and ethnographic research (Tomagová, Farský, Bóriková, & Zanovitová, 2016). In the Polish context, researchers concluded that

> Participation in various forms of educational activity broadens one's horizons. Seniors look to broaden their knowledge and circle of interests, drawing from a huge pool of opportunities. They often discover their passions and potential anew. To summarize, the participation of seniors in U3A classes is beneficial for the quality of their lives. U3A classes prevent isolation and result in positive changes of both an individual and social dimension and generates long-term benefits.
>
> (Mackowicz & Wnek-Gozdek, 2016: 196)

U3As are also commended for resolving the tensions arising from the push towards the productive use of one's free time and the pull of 'liberation' or 'well-earned rest' of retirement (Formosa, 2014). Indeed, when members are asked what they gain from their involvement in U3A activities, the first thing that they usually report is not generally related to the learning activities per se but to associated social outcomes such as making new friends and joining support groups (ibid., 2016).

However, the U3A movement has not escaped criticism. A consistent criticism levelled at U3As is that of elitism, as both survey and ethnographic data uncover a compounding class divide amongst membership bodies (Formosa, 2000, 2007, 2010). Although U3As offer no hindrances to admission, membership bodies tend to be exceedingly middle class. Indeed, whilst to middle-class elders joining U3As means going back to an arena in which they feel confident and self-assured of its outcome and development, working-class elders are apprehensive to join an organisation with such a 'heavy'-class baggage in its title. Patterson et al.'s (2016: 1598) qualitative study of a U3A in North-East England uncovered three exclusionary factors which acted as barriers for enrolment and participation—namely 'lack of knowledge about group presence and purpose (both locally and nationally), organisational name and location'. Reflecting other international studies, this study identified this U3A as a middle-class organisation that was mostly frequented by older persons with higher-than-average levels of educational attainment. U3As have also been criticised for including gender biases that work against the interests of both men and women

(Formosa, 2005). On the one hand, U3As tend to be characterised by a 'masculinist' discourse where women are silenced and made passive through their invisibility, an outnumbering of male over female tutors and a perception of older learners as a homogenous population which contributed towards a 'malestream' learning environment. On the other hand, the low percentage of male participants signals strongly that the organisation is not attractive to them. After all, most U3As include study units that generally reflect the interests of a female audience rather than conventional male interest in the physical and natural sciences. Another predicament is that U3As rarely include ethnic minorities in their membership bodies, even in multi-cultural cities such as Sydney and Auckland, nor do they generally cater for older persons experiencing physical and cognitive difficulties. In this respect, Ratana-Ubol and Richards' (2016) study of the U3A movement in Thailand demonstrated that by focusing on returning to learning, as U3As do, the movement overlooks the fact that many older persons outside the Global North (one may add the ethnic minorities in high-income cities) may have never attended any formal education when younger, and thus, may be more interested in basic literacy provision rather than what is generally offered. As they concluded,

> …as the Thailand example illustrates, this concept is sometimes resisted (or may not sufficiently inspire) in non-OECD or at least more traditional societies on the basis of cross-cultural differences as well as views of education directly linked to the tension between tradition and modernity (and associated rural-urban, poor-rich, and technological divides).
>
> (Ratana-Ubol & Richards, 2016: 99)

Due to the consistent emphasis that U3As create the 'conditions under which one becomes a Third Ager, a character in a new stage of life' where 'blanket phrases, which include "the Elderly", "Senior Citizens", "The Retired", and so on, have ceased to be appropriate', it is not surprising that the U3A movement has been 'accused at times of advocating "third Ageist" cultural and lifestyle priorities over economic and political issues, and for glorifying the positivity of the Third Age at the expense of the Fourth Age' (Katz, 2009: 153). Indeed, Wilińska (2012) found that the U3A movement in Poland serves as a vehicle for older persons to resist the negativity of their status and instead assert the meaningfulness of their age in the context of social attitudes to ageing:

> The results of this study indicate that rather than resisting ageist discourses, the U3A simply rejects the idea of old age. The U3A characterizes its members as exceptional people who have nothing in common with old people outside of the U3A. Therefore, the U3A plays only a minor role in changing the social circumstances of old people in Poland … 'If you are interested in ageing and older people in Poland, this is not a good place; walk the streets of the city, visit some care centres …'.
>
> (Wilińska, 2012: 290, 294)

The conclusions in Wilińska's (2012) study are reminiscent of Hazan's (1996) ethnography at the Cambridge U3A which elaborated upon the interpersonal discourses and rituals through which members reworked their identities as a kind of 'buffer zone' between midlife and old age:

...the members of the U3A distinctly referred to themselves as belonging not to the category of "old age" but to a category called the "third age". This category is defined as preceding "old age", but still as different from it...It was, in the eyes of members, a social buffer zone between past upward careers and social integration on one hand, and prospects of disengagement and deterioration on the other.

(Hazan, 1996: 33)

For Hazan (1996), the U3A experiment is one that encourages an anti-ageing and death-denying ambiance whereby the agency of members triumphs over the dark side of physical and socio-economic realities so prevalent in later life.

Book Outline

Four parts demarcate the book's chapters. The first part, 'The background context', includes two chapters titled 'Active ageing through lifelong learning: The University of the Third Age' (Marvin Formosa) and 'Origins and development: The Franco-phone model of Universities of the Third Age' (François Vellas). These chapters set the scene for the book's two subsequent parts by tracing the interface between 'active ageing' and 'lifelong learning in later life', and the origins of the University of the Third Age in Toulouse in 1973.

The book's second part, 'European Perspectives', includes nine chapters. Chapter three, '"An alternative ageing experience": An account and assessment of the University of the Third Age in the United Kingdom' (Keith Percy), illustrates how the U3A in the UK contributes significantly to positive and active ageing, taking older people out of their homes to join peers in physical, cognitive, expressive, imaginative, explorative, altruistic, interactive, creative and social activities. Chapter four, 'Be active through lifelong learning! The University of the Third Age in Iceland' (Hans Kristján Guðmundsson), underlines how the key message directed to the 50-plus generation by the Icelandic U3A movement is that, irrespective of age, it is never too late to rethink one's situation in life, embark on new pastimes, fulfil one's ambitions and act upon one's lifelong dreams and desires. Chapter five, 'The University of the Third Age in Italy: A dynamic, flexible, and accessible learning model' (Barbara Baschiera), discusses the contributions of U3As to lifelong learning for older adults in Italy, with specific reference to its educational, cultural and social impacts. Chapter six, 'Subsisting within public universities: Universities of the Third Age in Germany' (Bernhard Schmidt-Hertha), argues that although the structures which German U3As inhabit may be very specific to the surrounding geographical landscape, the popular didactical concepts reflect and mirror those present in other countries. Chapter seven, 'Third age learning for active ageing in Malta: Successes and limitations' (Marvin Formosa), reports on a multi-method study investigating the impact on Third Age Learning on active ageing, and which found that participation at the U3A impacted active ageing in three key ways—namely health, social inclusion and independence. Chapter eight, 'Late-life learning for social inclusion: Universities of the Third Age in Poland' (Jolanta Maćkowicz & Joanna Wnęk-Gozdek), includes an overview of ageing-related policy in Poland and its influence on older adult education, as well as the results of a research study on U3As which demonstrated the impact of seniors' participation on their physical, psychological and social well-being. Chapter nine, 'Universities of the Third Age: Learning opportunities in Russia' (Gulnara Minnigaleeva), documents how the role of U3As in Russia has been more that of engaging seniors in active lifestyles, assisting them to socialise with same-aged peers, and maintaining their mental health, rather than educating them for credential qualifications. Chapter ten, 'From university extension classrooms to universities of experience: The University of the Third Age in Spain' (Feliciano Villar), examines the historical, social and conceptual grounds which gave rise to the Spanish U3A movement, whilst highlighting its dual nature whereby a bottom-up model (university classrooms), based on lecture series organised by older people associations, coexists with a more top-down and academic approach (universities of experience), in which universities organise and offer degree-like programmes tailored to older students. The final chapter in the first part, 'Sweden's senior university: *Bildung* and fellowship' (Cecilia Bjursell), provides an overview of the Swedish Senior University movement, noting that there are 34 Senior Universities across the country, and that they are organised as independent associations but linked to the Swedish *Folkuniversitet* system, one of ten educational associations that exist in the Swedish *Folkbildning* organisation.

The book's third part, 'Asian-Pacific Perspectives', includes another nine chapters. Chapter twelve, 'The University of the Third Age movement in Australia: From statewide networking to community engagement' (Ainslee Lamb), describes the movement's origin and organisation in Australia, as well as its membership profile, curriculum approach and ongoing challenges. Chapter thirteen, 'The development and characteristics of Universities of the Third Age in mainland China' (Xinyi Zhao & Ernest Chui), provides an account of the historical development of elder learning in mainland China. It provides the policy framework, curriculum and student profile, teaching and learning strategies and the future challenges facing the U3A's movement in this vast country. Chapter fourteen, 'Third age learning in Hong Kong: The Elder Academy experience' (Maureen Tam), reports on the travails of the opportunities for Third Age Learning in Hong Kong, with special focus on the successes and limitations of Elder Academy network for older adult learning. Chapter fifteen, 'The University of the Third Age in Lebanon: Challenges, opportunities and prospects' (Maya Abi Chahine & Abla M. Sibai), documents the inspiring 'University for Seniors' programme, established in 2010 on the premises of the American University of Beirut, which presently provides adults aged 50-plus with the opportunity to impart knowledge and passions, share their life and professional experiences and remain intellectually energised and socially engaged. Chapter sixteen, 'Moving the needle on the University of the Third Age in Malaysia: Recent developments and prospects' (Tengku Aizan Hamid, Noor Syamilah Zakaria, Nur Aira Abd Rahim, Sen Tyng Chai & Siti Aisyah Nor Akahbar), records how the U3A programme in Malaysia renewed a focus on learning in later life and promoted a model where older persons themselves organise their own learning activities. Chapter seventeen, 'Universities of the Third Age in Aotearoa New Zealand' (Brian Findsen), explores the development of the U3A movement in Aotearoa, New Zealand, as a prominent initiative in older adult education provision in the wider sociocultural context of this bi-cultural country. Chapter eighteen, 'Third age education and the Senior University movement in South Korea' (Soo-Koung Jun and Karen Evans), outlines the origins, characteristics and structures of the Senior University in Korea and the perspectives that have guided its development. Its objective was to promote active ageing, and in doing so, to prioritise the least-advantaged group of Korean citizens. Chapter nineteen, 'From social welfare to educational gerontology: The Universities of the Third Age in Taiwan' (Shu-Hsin Kuo and Chin-Shan Huang), explains how the U3A movement in Taiwan arose to meet part of the challenges of population ageing in the country as it was thought that providing older adults with educational and learning opportunities will enable them to age actively and successfully. The final chapter in this third part, "Lifelong education' versus 'learning in later life': A University of the Third Age formula for the Thailand context?' (Cameron Richards, Jittra Makaphol and Thomas Kuan), explores the development of an alternative strategy by a traditional university wishing to create a new U3A centre in Thailand, through a local model which overlaps with both the Francophone and Anglophone U3A models, but as a community-based learning group supported by both a traditional university and the municipal resources.

The book's fourth part, 'Coda', which includes one chapter, 'Concluding remarks and reflections' (Marvin Formosa), brings this book to a close by serving as a beacon for further research studies on the U3A movement, whilst also highlighting both its contemporary and future challenges.

References

Åberg, P. (2016). Nonformal learning and well-being among older adults: Links between participation in Swedish study circles, feelings of well-being and social aspects of learning. *Educational Gerontology, 42*(6), 411–422.

Boudiny, K. (2013). 'Active ageing': From empty rhetoric to effective policy tool. *Ageing & Society, 33*(6), 1077–1098.

Boudiny, K., & Mortelmans, D. (2011). A critical perspective: Towards a broader understanding of 'active ageing'. *E-Journal of Applied Psychology, 7,* 8–14.

Boulton-Lewis, G., & Tam, M. (Eds.). (2012). *Active ageing, active learning: Issues and challenges.* Dordrecht, Netherlands: Springer.

de Palo, V., Limone, P., Monacis, L., Ceglie, F., & Sinatra, M. (2018). Enhancing e-learning in old age. *Australian Journal of Adult Learning, 58*(1), 88–109.

Estes, C. (1979). *The aging enterprise.* San Francisco: Jossey Bass.

Eurostat. (2018). *Participation rate in education and training (last 4 weeks) by type, sex and age.* http://appsso.eurostat.ec.europa.eu/nui/submitViewTableAction.do. Accessed September 24, 2018.

Findsen, B., & Formosa, M. (2011). *Lifelong learning in later life: A handbook on older adult learning.* Rotterdam, Netherlands: Sense Publishers.

Findsen, B., & Formosa, M. (2016a). Introduction. In B. Findsen & M. Formosa (Eds.), *International perspectives on older adult education: Research, policies, practices* (pp. 1–9). Cham, Switzerland: Springer.

Findsen, B., & Formosa, M. (2016b). Concluding remarks. In B. Findsen & M. Formosa (Eds.), *International perspectives on older adult education: Research, policies, practices* (pp. 507–519). Cham, Switzerland: Springer.

Findsen, B., Mcewan, B., & Mccullough, S. (2011). Later life learning for adults in Scotland: Tracking the engagement with and impact of learning for working-class men and women. *International Journal of Lifelong Education, 30*(4), 527–547.

Formosa, M. (2000). Older adult education in a Maltese University of the Third Age: A critical perspective. *Education and Ageing, 15*(3), 315–339.

Formosa, M. (2005). Feminism and critical educational gerontology: An agenda for good practice. *Ageing International, 30*(4), 396–411.

Formosa, M. (2007). A Bourdieusian interpretation of the University of the Third Age in Malta. *Journal of Maltese Education Research, 4*(2), 1–16.

Formosa, M. (2010). Universities of the Third Age: A rationale for transformative education in later life. *Journal of Transformative Education, 8*(3), 197–219.

Formosa, M. (2011). Critical educational gerontology: A third statement of principles. *International Journal of Education and Ageing, 2*(1), 317–332.

Formosa, M. (2012). European Union policy on older adult learning: A critical commentary. *Journal of Aging and Social Policy, 24*(4), 384–399.

Formosa, M. (2014). Four decades of Universities of the Third Age: Past, present, and future. *Ageing & Society, 34*(1), 42–66.

Formosa, M. (2016). Malta. In B. Findsen & M. Formosa (Eds.), *International perspectives on older adult education: Research, policies, practices* (pp. 261–272). Cham, Switzerland: Springer.

Formosa, M. (2017). Responding to the Active Ageing Index: Innovations in active ageing policies in Malta. *Journal of Population Ageing, 10*(1), 87–99.

Formosa, M. (2018). National policies for healthy ageing in Malta: Achievements and limitations. *Healthy Aging Research, 7*(1), 1–6.

Funnell, I. (2017). The development of U3As & the benefits to members. *Roczniki Nauk. Społecznych, 2,* 119–137.

Glendenning, F., & Battersby, D. (1990). Why we need educational gerontology and education for older adults: A statement of first principles. In F. Glendenning & K. Percy (Eds.), *Ageing, education and society: Readings in educational gerontology* (pp. 219–231). Keele: Association for Educational Gerontology.

González-Palau, F., Franco, M., Bamidis, P., Losada, R., Parra, E., Papageorgiou, S. G., et al. (2014). The effects of a computer-based cognitive and physical training program in a healthy and mildly cognitive impaired aging sample. *Aging & Mental Health, 18*(7), 838–846.

Guillemard, A.-M. (1981). Old age, retirement and the social class structure: Towards an analysis of the structural dynamics of the latter stage of life. In T. K. Hareven (Ed.), *Dying and life course transitions* (pp. 221–243). New York: Guilford Press.

Hachem, H., Nikkola, E., & Zaidan, A. (2017). The case of educational gerontology in Lebanon: A harbinger of empowerment, emancipation and social change? *International Journal of Lifelong Education, 36*(6), 713–729.

Hafford-Letchfield, T. (2014). Critical educational gerontology: What has it got to offer social work with older people? *European Journal of Social Work, 17*(3), 433–446.

Hafford-Letchfield, T. (2016). *Learning in later life: Challenges for social work and social care.* London: Routledge.

Harris, R., & Ramos, C. (2013). Building career capital through further study in Australia and Singapore. *International Journal of Lifelong Education, 32*(5), 620–638.

Havighurst, R. (1963). Successful ageing. In R. Williams, C. Tibbitts, & W. Donahue (Eds.), *Process of ageing* (Vol. I, pp. 299–320). New York: Atherton.

Hazan, H. (1996). *From first principles: An experiment in Ageing.* Westport, CT: Bergin & Garvey.

International Longevity Centre Brazil. (2015). *Active ageing: A policy framework in response to the longevity revolution.* Brazil: International Longevity Centre Brazil.

Jenkins, A., & Mostafa, T. (2015). The effects of learning on wellbeing for older adults in England. *Ageing & Society, 35*(10), 2053–2070.

Kalache, A. (1999). Active ageing makes the difference. *Bulletin of the World Health Organization, 77*, 299.

Katz, S. (2009). *Cultural aging: Life course, lifestyle, and senior words.* Toronto: University of Toronto Press.

Kimberley, H., Golding, B., & Simons, B. (2016). The company of others: Generating knowhow in later life. *International Journal of Lifelong Education, 35*(5), 509–521.

Laslett, P. (1989). *A fresh map of life: The emergence of the third age.* London: Macmillan Press.

Lido, C., Osborne, M., Livingston, M., Thakuriah, P., & Sila-Nowicka, K. (2016). Older learning engagement in the modern city. *International Journal of Lifelong Education, 35*(5), 490–508.

Mackowicz, J., & Wnek-Gozdek, J. (2016). It's never too late to learn"—How does the Polish U3A change the quality of life for seniors. *Educational Gerontology, 42*(3), 186–197.

Maginess, T. (2017). *Enhancing the wellbeing and wisdom of older learners. A co-researched paradigm.* London: Routledge.

Maniecka-Bryła, I., Gajewska, O., Burzynska, M., & Byrla, M. (2013). Factors associated with self-rated health (SRH) of a University of the Third Age. *Archives of Gerontology and Geriatrics, 57*, 156–161.

McClusky, H. Y. (1974). Education for ageing: The scope of the field and perspectives for the future. In S. Grabowski & W. D. Mason (Eds.), *Learning for ageing* (pp. 324–355). Washington, D.C.: Adult Education Association of the USA.

Mercken, C. (2010). *Education in an ageing society.* Sittard: Drukkerij.

Midwinter, E. (1984). Universities of the Third Age: English version. In E. Midwinter (Ed.), *Mutual Aid Universities* (pp. 3–19). Kent: Croom Helm Ltd.

Moody, H. R. (1990). Education and the life cycle. In R. H. Sherron & B. Lumsden (Eds.), *Introduction to educational gerontology* (pp. 23–39). Washington, DC: Hemisphere.

Narushima, M., Liu, J., & Diestelkamp, N. (2018). Lifelong learning in active ageing discourse: Its conserving effect on health, wellbeing and vulnerability. *Ageing & Society, 38*(4), 651–675.

Niedzielska, E., Guszkowska, M., Kozdroń, E., Leś, A., Krynicki, B., & Piotrowska, J. (2017). Quality of life and its correlates in students of a University of the Third Age. *Polish Journal of Sport Tourism, 24*, 35–41.

O'Donnell, K. (2006). *Adult education participation in 2004–2005.* Washington DC: U.S. Department of Education. National Center for Education Statistics.

Patterson, R., Moffatt, S., Smith, M., Scott, J., McLoughghlin, C., Bell, J., et al. (2016). Exploring social inclusivity within the University of the Third Age (U3A): A model of collaborative research. *Ageing & Society, 36*(8), 1580–1603.

Paz, A., Doron, I., & Tur-Sinai, A. (2018). Gender, aging, and the economics of "active aging": Setting a new research agenda. *Journal of Women & Aging, 30*(3), 184–203.

Percy, K. (1990). The future of educational gerontology: A second statement of first principles. In F. Glendenning & K. Percy (Eds.), *Ageing, education and society: Readings in educational gerontology* (pp. 232–239). Keele: Association for Educational Gerontology.

Phillipson, C. (1982). *Capitalism and the construction of old age.* London: Macmillan.

Radcliffe, D. (1984). The international perspective of U3As. In E. Midwinter (Ed.), *Mutual Aid Universities* (pp. 61–71). Kent: Croom Helm Ltd.

Ratana-Ubol, A., & Richards, C. (2016). Third age learning: Adapting the idea to a Thailand context of lifelong learning. *International Journal of Lifelong Learning, 35*(1), 86–101.

Salter, B., & Salter, C. (2018). The politics of ageing: Health consumers, markets and hegemonic challenge. *Sociology of Health & Illness, 40*(6), 1069–1086.

Swindell, R. F. (2011). Successful ageing and international approaches to later-life learning. In G. Boulton-Lewis & M. Tam (Eds.), *Active ageing, active learning* (pp. 35–65). New York: Springer.

Talmage, C. A., Lacher, R. G., Pstross, M., et al. (2015). Captivating lifelong learners in the third age: Lessons learned from a university based institute. *Adult Education Quarterly, 65*(3), 232–249.

Talmage, C. A., Mark, R., Slowey, M., & Knopf, R. C. (2016). Age Friendly Universities and engagement with older adults: Moving from principles to practice. *International Journal of Lifelong Education, 35*(5), 537–554.

Teaching Study Department Guangzhou Elderly University. (2017). On history and development of Universities of the Third Age in China. *Roczniki Nauk Społecznych, 2,* 101–118. https://www.ceeol.com/search/article-detail?id=613566. Accessed September 2, 2018.

Tomagová, M., Farský, I., Bóriková, I., & Zanovitová, M. (2016). Selected indicators of mental health in the elderly. The participants at the University of the Third Age. *Central European Journal of Nursing and Midwifery, 7*(2), 437–443.

U3A Alliance Australia. (2017). *U3A Alliance Australia.* http://www.u3aaa.org/. Accessed November 22, 2018.

University of the Third Age. (2018). *Third Age Trust/University of the Third Age.* https://www.u3a.org.uk. Accessed March 2, 2018.

van Malderen, L., Mets, T., De Vriendt, P., & Gorus, E. (2013). The active ageing-concept translate to residential long-term care. *Quality of Life Research, 22*(5), 929–937.

Vellas, P. (1997). Genesis and aims of the Universities of the Third Age. *European Network Bulletin, 1,* 9–12.

Villar, F., & Celdrán, M. (2013). Learning in later life: Participation in formal, non-formal and informal activities in a nationally representative Spanish sample. *European Journal of Ageing, 10*(2), 135–144.

von Humboldt, S. (2016). *Conceptual and methodological issues on the adjustment to aging. Perspectives on aging well.* New York: Spinger.

Walker, A. (1980). The social creation of poverty and dependency in old age. *Journal of Social Policy, 9*(1), 49–75.

Walker, A. (1981). Towards a political economy of old age. *Ageing & Society, 1*(1), 73–94.

Walker, A. (2009). Commentary: The emergence and application of active aging in Europe. *Journal of Ageing & Social Policy, 21*(1), 75–93.

Wang, R., De Donder, L., De Backer, F., Triquet, K., Shihua, L., Honghui, P., et al. (2018). Exploring the association of learning participation with the quality of life of older Chinese adults: A mixed methods approach. *Educational Gerontology, 44*(5–6), 378–390.

Wilińska, M. (2012). Is there a place for an ageing subject? Stories of ageing at the University of the Third Age in Poland. *Sociology, 46*(2), 290–305.

Withnall, A. (2010). *Improving learning in later life.* London: Routledge.

Withnall, A. (2016). United Kingdom. In B. Findsen & M. Formosa (Eds.), *International perspectives on older adult education: Research, policies, practices* (pp. 467–480). New York: Springer.

World Health Organization. (1994). *Health for all: Updated targets.* Copenhagen: World Health Organization.

World Health Organization. (2002). *Active ageing: A policy framework.* Geneva: World Health Organization.

Zielinska-Wieczkowska, H., Ciemnoczołowski, W., Kedziora-Kornatowska, W., & Muszali, M. (2013). The sense of coherence (SOC) as an important determinant of life satisfaction, based on own research, and exemplified by the students of University of the Third Age (U3A). *Archives of Gerontology and Geriatrics, 54,* 238–241.

Marvin Formosa is Associate Professor of Gerontology at the University of Malta where he is Head of the Department of Gerontology and Dementia Studies, Faculty for Social Wellbeing, and contributes to teaching on active ageing, transformative ageing policy, and educational gerontology. He holds the posts of Chairperson of the National Commission for Active Ageing (Malta), Rector's Delegate for the University of the Third Age (Malta), and Director of the International Institute on Ageing, United Nations, Malta (INIA). He directed a number of international training programmes in gerontology, geriatrics and dementia care in the Philippines, China, India, Turkey,

Malaysia, Belarus, Kenya, Argentina, Azerbaijan and the Russian Federation. He has published extensively across a range of interests, most notably on active ageing, critical gerontology, Universities of the Third Age and older adult learning. Recent publications included *International perspectives on older adult education* (with Brian Findsen, 2016), *Population ageing in Turkey* (with Yeşim Gökçe Kutsal, 2017), and *Active and healthy ageing: Gerontological and geriatric inquiries* (2018). He holds the posts of Editor-in-Chief of the International Journal on Ageing in Developing Countries, Country Team Leader (Malta) of the Survey of Health, Ageing, and Retirement in Europe (SHARE), and President of the Maltese Association of Gerontology and Geriatrics.

Chapter 2
Origins and Development: The Francophone Model of Universities of the Third Age

François Vellas

Introduction

Universities of the Third Age (U3As) constitute organisations that provide education and related activities to retired people (Formosa, 2014). Most retired people are considered to be in the 'third age' of life, because it is the age following the life stage of full-time employment and parental responsibilities, known as the second age (Laslett, 1989). Although there is no commonly accepted age threshold for the start of the Third Age, this life stage is generally considered to begin at around 60 years of age (Midwinter, 2005). The fast development of U3As throughout the world is testimony to the growing importance of the older generation within modern society. Nowadays, many U3As form part of the International Association of Universities of the Third Age (AIUTA). Despite this, demographic ageing is too often perceived negatively, and older people are generally considered to be a burden on society, and on public financial budgets; a change of attitude is required to achieve an inclusive society with more solidarity and cooperation with the older generation. The first University of the Third Age (U3A) was founded in 1973 at the University of Toulouse by Pierre Vellas (Philibert, 1984; Radcliffe, 1984; Swindell & Thompson, 1995; Picton & Lidgard, 1997). Vellas (1997) had two key objectives: to meet the expectations of older people in good health who were looking for post-retirement activities and to enable the drive of university outreach to older learners. He thus sought to meet these two expectations by creating a learning institution that welcomed older students into a mainstream university. In Formosa's words,

Vellas held that the goal of the U3A was to investigate, without any preconceived notions, how universities could improve the quality of life of older persons who, as demographic statistics at that time suggested, were becoming more numerous and whose socioeconomic conditions were often in a deplorable condition. Four major objectives were formulated for this new educational enterprise. These included raising the quality of life of older people,

F. Vellas (✉)
University of Toulouse Capitole, Toulouse, France
e-mail: fvellas@gmail.com

© Springer Nature Switzerland AG 2019
M. Formosa (ed.), *The University of the Third Age and Active Ageing*, International Perspectives on Aging 23, https://doi.org/10.1007/978-3-030-21515-6_2

realizing a permanent educational program for older people in close relations with younger peers, coordinating gerontological research programs, and last, but not least, realizing initial and permanent education programs in gerontology.

(Formosa, 2010: 115)

Indeed, Vellas did not seek to create a university that prioritises accreditation, since older learners are not generally looking to re-enter the labour market, but rather for an opportunity to engage in learning as the end goal (Findsen & Formosa, 2011). Indeed, older learners generally wish to acquire information and skills that they did not have time to acquire when their life consisted of more obligations. Vellas' initiative was a success, and U3As are nowadays found in many countries and in all continents. This chapter traces the origins and development of the Francophone U3A model in Toulouse, noting its close affiliation to traditional universities, and contrasting its modus operandi with U3As following the British model. The chapter also commemorates the work being done by the French Association of Universities of the Third Age and International Association of the Universities of the Third Age, the overlap between U3As and the European Federation of Older Students, as well as the range of European projects that U3As have recently been involved in.

The Toulouse U3A

Originally, the Toulouse U3A offered a course on gerontology within the University of Toulouse's Faculty of Social Sciences. Lectures were exclusively for persons above the statutory retirement age of the time. Learning programmes focused on memory expansion and activities that improve diet and general health. Building on the success of the first programme of studies, Vellas initiated new courses, such as gymnastics and languages, and eventually set up a new department at the University of Toulouse's Faculty of Social Sciences as a branch of Toulouse University I (presently named *Toulouse Capitole*), and as part of the Department for Continuing Education. Looking back on the launch of this innovative movement, Formosa argued that in retrospect,

...there was nothing exceptional about this programme apart from the fact that a section of a large provincial university had taken an interest in ageing and decided to enlist the resources of the university in programmes for senior citizens. Yet, the Toulouse initiative struck a rich vein of motivation so that just three years later U3As were already established in Belgium, Switzerland, Poland, Italy, Spain, and Quebec in Canada...The U3A phenomenon struck a rich vein of motivation because retirees perceived such centres as offering them the possibility to continue engaging in physical and cognitive activities even beyond retirement, and to keep abreast of physical, psychological, and social changes occurring in later life...This may be because the U3A movement was in marked contrast to the tradition of centralised educational management, and provided an opportunity to sow the first seeds of educational innovation and reform.

(Formosa, 2014: 44)

Presently, the Toulouse U3A benefits from all the resources of the University of Toulouse, including having professors from various disciplines to deliver lectures and the use of information and communication technology equipment to teach foreign languages. As a result, the Toulouse U3A does not require additional financial resources with respect to the location of suitable premises or the employment of academic lecturers, and a secretariat pool is available on a voluntary basis. For this reason, U3A members are only required to pay a nominal fee.

During the 2015–2016 academic year, classes held an average of 60 participants, although lectures on languages generally attracted over 400 students. The themes of the lectures nowadays are diverse—ranging from Tai chi, languages, and information and communications technology to history, philosophy, art, religion, law, economics and natural sciences. Each class lasts 1 h 30 min, held twice a week for 42 consecutive weeks during the academic year. The Toulouse U3A also offers a monthly schedule of guided tours. The guide in charge of the tours is a Professor of Art History, paid €110 for each tour. The tours consist of a series of afternoon walks and therefore do not require the use of transport and do not offer meals. They usually attract between 25 and 30 participants and are free except for entrance fees to museums and monuments, which range between €2 and €10. Optional tours to surrounding towns are also organised about three or four times a year. The towns are usually located less than 150 km from Toulouse and include Albi, Rodez, Cahors, Foix, Carcassonne, Condom, Castres, and Cordes sur Ciel. These one-day tours—which include transport, lunch and dinner—are again organised by the Professor of History of Art in collaboration with a local transport company. Costs range between €60 and €100. Health education is a very important topic for U3A members, and similar to what happens in other French U3As, the Toulouse U3A has strong links with the Faculty of Medicine and the university's teaching hospitals. The latter offers lectures increasing awareness of major illnesses relating to later life such as Alzheimer's and dementia. Specific physical exercises are provided by the U3A to assist its members to live healthily, and gym sessions are also provided by dedicated instructors. Every year, the Toulouse U3A offers its members the opportunity to join the AIUTA delegation in its annual conference in countries as diverse as Poland, China, Japan and Brazil. Typically, air travel is paid by the U3A members themselves, whilst the trip is organised by a U3A member in cooperation with the parent U3A in the foreign country on a voluntary basis.

U3A Affiliations to Universities

Older persons from rural areas, who wished to enrol in a U3A, inspired a new category of U3As seeking affiliation with mainstream universities. Two such examples are the Rodez U3A which is affiliated to the University of Toulouse and the Orleans U3A which is allied with the University of Orleans. The Rodez (or Rouergue) U3A—also known as a University of Free-Time (or Leisure)—is a non-profit association launched by Monique Viala in 2006 and directed by François Yence since 2012. Its aim is to develop cultural and educational activities to facilitate access to knowledge for older persons as well as to promote social ties and exchanges between members, and between the University and the population of Rodez. Although this association belongs to the network of U3As, in theory it is open to everybody, irrespective of age, but in practice most of its members are over 60 years of age. As an association, it bears similarities to some of the Australian U3As such as the one in Melbourne. Its directors and committee members—elected by members in an Annual General Meeting—are responsible for operational decisions. The Orleans University of the Third Age was founded in 1977, to be later renamed the Orleans' University of Free Time (UFT). The Orleans UFT forms part of the Faculty of Arts and Literature's lifelong education programme within the University of Orleans. However, the Orleans UFT is financially independent, with its regulations set by the University of Orleans' Board of Directors. Since there are as many as nine UFTs within 200 km of each other, UFT members attend lectures in different towns and cities. This allows members to access a greater variety of lectures, meet other members not living in their locality and enjoy morning trips together.

Continental European Trials

The U3A Francophone model has been implemented in many a European university. One such example is the U3A at Comenius University in Bratislava which forms part of the Department for Continuing Education. Since its establishment in 1990, the Comenius U3A organised learning programmes at the main campus of Comenius University, charging its members a nominal fee for the service. In 2005, the U3A office was moved to the newly-built Centre for Continuing Education, which is independently financed and managed. As a result, the Centre became responsible for all of the U3A's expenditure, with the Comenius University's Rector being designated as U3A President, and thus, being made responsible for the signing of all European Union project applications. The Centre for Continuing Education Director assumes full responsibility of the U3A's work, but always on behalf of the Rector. A committee chaired by the Rector, in cooperation with the Vice-Rector for the education and the Director of the Centre, is responsible for the setting of the learning programme. The U3A's curriculum is not solely based on traditional academic and educational interests, but also incorporates health-related, cultural and leisure activities.

The British U3A Model: U3As Registered as Non-governmental Organisations

U3As in Britain are mainly registered as associations and non-governmental organisations (Huang, 2006, Manheimer, Snodgrass, & Moskow-Mckenzie, 1995; Midwinter, 1984, 2004; Withnall, 2016). The most common legal framework for British U3As is that of an 'association' status, termed a 'charitable organisation' or 'non-governmental organisation'. There is no law in Britain regulating older adult education or lifelong learning. The state has also little involvement in the British U3A movement, whether at governmental or local level, and U3As are mostly independent associations. To cite Formosa,

> The hallmarks of British U3As include their sturdy independence and anti-authoritarian stance…Aspiring to instigate a democratic movement that enriches the lives of older adults through the development of a range of learning, action, and reflection opportunities, British U3As declined to form part of "the official, state-founded, established structure with its professional teachers and administers" (Laslett, 1989: 174). Instead of developing into campus-based organisations (although the Lancaster and London U3As were notable exceptions), British U3As…stressed the oppressive and ineffectual nature of institutionalised education, on the basis that whilst institutionalised learning undermines people's confidence and undermines their capacity to solve problems, the so-called experts (that is, teachers) tend to self-select themselves and act as gate-keepers as what should be learnt in life.
>
> (Formosa, 2014: 45)

Nevertheless, British U3As group themselves under a national representative body—namely the Third Age Trust. From the Cambridge U3A's website one learns that while this association offers a wide variety of courses (more than 350), only a handful of academic lectures are in place (University of the Third Age Cambridge, n.d.). Although the Cambridge U3A offers fewer academic lectures than Toulouse U3A (once, sometimes twice, a week), the number of learning programmes is larger. Although these courses are held in small groups and may not be under the tutelage of professors, they are often related to academic subjects such as history, science and economics. Among the learning programmes made available by the Cambridge U3A, 43% belong to this category, generally offered for one or two terms of three months each, whereas other activities such as fitness are year-round.

Francophone and Anglophone Comparisons

In France, U3As are grouped within the French Association of Universities of the Third Age (*Union Française des Universités Tous Âges*) (UFUTA), a non-profit organisation created in 1981 by René Frentz from Nancy U3A, with its head office located at the Centre of Links, Studies, Information and Research on the Problems of Elderly People (CLEIRPPA) in Paris. In 2018, UFUTA comprised of 35 member organisations boasting a total of 65,000 members. The link between UFUTA and the higher education system is evident through its willingness to promote research and geragogy in older adult education. Indeed, one of its main aims is to create and coordinate collective programmes on teaching and learning in later life; research in older adult learning; and advocacy to promote the development of lifelong learning. In Britain, U3As are members of *The Third Age Trust* which was registered as a charity in 1983, and which is managed by a National Executive Committee consisting of a chairperson, three other officers, and 12 representatives from all regions (Third Age Trust, n.d.). The Third Age Trust is funded by annual subscriptions from member

U3As, with the occasional grant for specific projects. It includes seven full-time and seven part-time staff located in Bromley. In 2018, there were 1028 U3As in the UK with over 400,000 members. The Third Age Trust has recently revived its research committee, aiming to identify U3As' research activities and encourage more research on the U3A's gender balance, size distribution and membership composition. In an insightful comparison of French and British U3A movements and arrangements, Formosa underlined that

> ...the French and British U3As models were a product of particular socio-political contexts...the French U3As had legislation on their side, together with a conviction that education for older adults was a necessary part of universities responsibilities, so that it is not surprising that their key distinguishing feature is their association with established universities and government departments responsible for the welfare of older people. On the other hand, the proposals of the working party on *Education and Older Adults* set up by the British Universities Council for Adult and Continuing Education in 1982 elicited a poor response from universities and political parties...British U3As had no other alternative but to adopt the principles of self-help learning, as evidenced from their strong will to remain free of universities, autonomous from local authorities, and downplay traditional credentials.
>
> (Formosa, 2014: 46)

It is noteworthy that the French UFUTA and the British Third Age Trust have much in common. For instance, they both stress the concept of learning as an end-in-itself and reject the idea of qualifications or formal certifications. However, they also differ in many ways.

Link with Universities. In France, the term 'University' in the U3A title implies a link with higher educational institutions. This link requires membership with the UFUTA, and without it, an organisation would not be able to use title of U3A. Indeed, to qualify for membership with the UFUTA, a U3A must be a department of or a unit in a higher education establishment (an affiliation is also acceptable). This is not the case with the Third Age Trust, where the term 'University' in the U3A title is used in its original medieval sense of people coming together to share and pursue learning in all its forms, without necessarily having a formal link with a higher educational institution. In Britain, U3A activities may take place in community centres or even in members' own homes. This model is also replicated in some of the other Commonwealth countries, such as Australia and New Zealand.

Geragogy and research. The French U3A model is a top-down model where scientific research is carried out by experts, in order to determine the best programmes for older adults. On the other hand, the British U3A model follows a bottom-up approach where learning programmes are democratically decided on, according to the needs, interests and preferences of the members.

Lobbying and official recognition of U3As. The aims of the UFUTA include advocacy work with public authorities in order to obtain the official recognition of its member organisations. In France, education—including that which is lifelong—is government-funded. Protests, held in May of 1968, resulted in amendments to the Higher Education Act which stated that 'French Universities must be open to all categories of the population', so that in 1971 a new law forced French Universities to fund adult and older adult education as part of the country's national lifelong learning policy. In addition, French municipalities have the right to fund associations serving the public good (*associations d'utilité publique*); meaning that U3As in France can receive financial support from the government, either as part of funding measures towards tertiary education, or as financial subsidies to non-governmental organisations. There is, therefore, a strong incentive for the government to acknowledge U3As in an official capacity and to maintain or increase their financial support. Hence, lobbying public authorities is one of UFUTA's main functions.

International Association of the Universities of the Third Age (AIUTA)

International cooperation is one of the U3A's most important objectives; the International Association of the Universities of the Third Age (AIUTA) brings together U3As from all continents around the world (AIUTA, n.d.). AIUTA encouraged U3As to enrol as members as a means of being in touch with other U3As so that your own development of programmes for senior citizens can be informed by the actions, reflections and innovations of other U3As; in order to come together from time to time with colleagues working in different contexts to share ideas and experiences; to situate local activity within a wider context through AIUTA's international and global focus; to both receive help and act as a source of support for U3As in their contacts with institutions and organisations of all types; to take part in the global development of awareness of the needs and contributions of senior citizens through a dedicated virtual network of contacts across the world; to use AIUTA's international network to establish exchanges between U3As in different countries; and to contribute by offering ideas and initiatives to other institutions active among senior citizens (ibid.). Through AIUTA's work, older learners are able to interact with others from different countries by participating in study tours and building cross-national friendships. International exposure also provides lecturers with greater global insight into the teaching of seniors and in researching gerontological issues. Programmes involving U3As benefit society in the fields of medicine and economic development based on the 'silver' economy ethos. Membership in AIUTA is also open to organisations pursuing the same objectives as U3As such as Universities of Free Time and Inter-Age Universities. Hence, AIUTA serves as a federation of U3As that includes national federations and other partner institutions representing U3As from various countries such as France, Spain, Portugal, Italy, Sweden, Switzerland, Slovakia, the UK, Poland, Croatia, the Czech Republic, Romania, Argentina, Brazil, Canada, Chile, China, Malta, Japan, Lebanon, Australia, Mauritius, Senegal, Singapore and Tunisia—among others.

Foundation and aims. Pierre Vellas founded AIUTA in 1975, with the objective of creating a network of different U3As in Europe and around the world; this was planned to focus on the following areas: teaching through conferences, courses, workshops and study visits, research and community services for older persons. At present, its aims include:

- linking U3As together so that innovations spread from one centre to another;
- convening cross-national members of U3As to share ideas and experiences;
- providing local U3As with an international dimension through AIUTA;
- supporting local U3As to interact with foreign counterparts;
- taking part in developing global awareness of the opportunities and issues concerning older persons through a virtual network of U3As;
- using AIUTA's international network to establish exchanges with U3As of different countries; and
- sharing ideas between organisations working for the benefits of older adults.

Cooperation. Until 1990, U3As in AIUTA developed their own research projects in partnership with each other under bilateral or multilateral agreements. In 2007, these partnerships were replaced by a system of enlarged networking, with projects conducted by at least three U3As from different European countries. These new European programmes created new opportunities for research projects involving several U3As within AIUTA. In January 2011, the author was elected President of AIUTA for a four-year term (2011–2014), and re-elected again in 2014 for an additional four-year term (2014–2018). During these two terms, AIUTA had a renewed focus, as could be witnessed by the increase in number of international conferences and research projects that it organised.

Strategic action. AIUTA's Education Committee advocates the need to improve U3As by providing advice on legal aspects and constitution; academic and pedagogical issues; technical and operational issues; and monitoring and educational standards.

Research. The main actions implemented by AIUTA's Research Committee include advice on how to design research projects and complete application forms for European funding, as well as how to develop international U3A cooperation under common research programmes. Moreover, AIUTA has published a number of research papers on senior tourism, seniors and the arts, seniors and pilgrimage, innovation in retirement homes, world observatory on senior tourism, creating a U3A, comparisons between the French and British U3A models, U3A curricula and the role of partner institutions in the development of the U3As, among others.

Networking. AIUTA offers institutional membership to U3As around the world as well as to national associations of U3As. It includes U3A members from Africa, Asia, Europe, the Middle East, and North and South America—many of which have been operating for many years. Moreover, AIUTA grants individual memberships to experts on older and Third Age education to strengthen its research and educational ideas.

European Federation of Older Students (EFOS)

The European Federation of Older Students in universities (EFOS) was founded in 1990 in Bressanone, Italy by Leopold Auinger (EFOS, n.d.). The aim of EFOS constitutes in (i) promoting high-level education of older people alongside younger students or at special academies/universities for older people; (ii) fostering of joint projects for older students throughout Europe; (iii) fostering and securing the access of older people, including those without formal qualifications, to education at the highest level; (iv) representing the interests of academic education for older people in politics and society; (v) fostering of autonomous national organisations with similar goals; (vi) identifying possibilities of using the skills and expertise of older students for the benefit of science and society; and (vi) cooperating with other international organisations that support lifelong learning (ibid.). Since its foundation, EFOS has organised biannual international meetings in various European countries. The association is affiliated with the United Nations in Vienna (a representative of EFOS is a member of the United Nations Committee on Ageing), as well as with AIUTA. The cooperation between European universities and senior university students is based on exchanging information on continuing education for older learners, exchanging life experiences between older students, supporting study programmes for older students, and conducting research on the topics of older adult education within the European context. EFOS includes members from the following countries: Austria, Czech Republic, Germany, Great Britain, the Netherlands, Poland, Slovakia, Spain, Sweden and Switzerland—most of whom are involved in new projects such as the Erasmus+ and EduSenNet. In May 2016, AIUTA and EFOS signed a *Memorandum of Understanding* which confirmed the two organisations' common interest in older adult learning. In brief, AIUTA and EFOS have agreed to collaborate towards the following goals:

- cooperation in the fulfilment of international events supporting the development of learning programmes, strategies and policies directed towards older people;
- cooperation in providing information in connection with the active participation of the public and in ensuring the awareness and active participation of governments and the European Commission in support of the further education of older people;
- cooperation in promoting and implementing research programmes and projects for older people; and
- cooperation in the exchange of information and other activities leading to the fulfilment of the memorandum's mission.

European Union Programmes Involving U3As

Erasmus+ is a European Union programme that aims to boost skills and employability in Europe, as well as modernising education, training and youth work (Formosa, 2012a, 2012b). It encourages cooperation and transnational partnerships between

European partners by encouraging countries to join together in tenders for individual Erasmus projects. Presently, Erasmus+ (2014–2020) has a €14.7 billion framework programmed for potential projects on education, training, youth, sports, as well as lifelong learning programmes (Erasmus, Leonardo da Vinci, Comenius, Grundtvig). AIUTA was successful in securing Erasmus+ funds through the Grundtvig stream. For example, a project focusing on proverbs, traditions and language was conducted by the University of Namur (Belgium) in collaboration with Barcelona U3A (Spain), Vannes U3A (France) and EFOS. Other past projects included:

- *The EFOSEC project* which aimed to create better conditions and possibilities for the research of teaching and learning needs for older persons, promote older adult education and training systems, and disseminate information about educational possibilities through thematic networks, the EFOS homepage and through online networks;
- *The VECU project* which developed methods for older people by which to pass on the cultural tradition of their country to people of another culture (intercultural) and to the younger generation (intergenerational); and
- *The Educational Senior Network* (*EduSenNet*) which summarised older adults' experiences and study possibilities for learning, thus encouraging older adults to develop their interest in learning through participation in activities designed to promote and extend their knowledge and skills.

Conclusion

AIUTA's future plans include the implementation of recommendations made past and present European Union research projects; reinforcing the U3As European network; consolidating a network with Asia Pacific Alliance U3As; lobbying for Third Age education in the European Union and United Nations; involving more U3As in research programmes; setting up the World Observatory of Senior Tourism (WOST); and creating centres of interest for research activities for both young and senior researchers from the various U3As, promoting intergenerational exchanges. Key goals guiding future AIUTA actions include:

- Improving the exchange and cooperation between AIUTA member institutions. AIUTA is proposing to set up 'a bank of knowledge for seniors and their U3As' which will be useful when organising symposiums, conferences and bilateral exchanges between U3As on an international scale.
- Addressing climate change, in particular in small islands such as the ones located in the Pacific, but also in other regions in coastal areas. U3As could assist government and non-governmental organisations with expertise from its lectureship and membership body. A forum on U3As' contribution to mitigate against climate change could contribute to these issues.
- Setting an International Senior Tourism Sustainable Observatory (WOST) that aids AIUTA in promoting tourism among older persons. The Observatory will provide

sustainable solutions to assist older persons in their travels, bringing about wider health benefits and social cohesion at the destination countries.

• Promoting intergenerational applied research and teaching, and geragogic practices for U3A members. This warrants a win-win strategy that is beneficial for all—namely senior students, partner institutions, the scientific community, AIUTA and its partner U3A associations and the regions covered by these networks.

AIUTA's ultimate goal is to enhance the well-being of older people around the world, by improving the acquisition and transfer of knowledge through global productive partnerships with other U3As associations, such as the Asia Pacific Alliance of U3As and the U3A Network Queensland, among others. It is positive to note that the U3A Asia Pacific Alliance (APA) is progressing remarkably well considering it operates informally with U3As in several different countries, that in turn offer to organise annual conferences. There is no doubt that the U3A movement is thriving and that the best is yet to come!

References

European Federation of Older Students. (n.d.). *European Federation of Older Students*. http://www. efos-europa.eu/. Accessed May 8, 2018.

Findsen, B., & Formosa, M. (2011). *Lifelong learning in later life: A handbook on older adult learning*. Rotterdam, Netherlands: Sense Publishers.

Formosa, M. (2010). Universities of the Third Age: A rationale for transformative education in later life. *Journal of Transformative Education, 8*(3), 197–219.

Formosa, M. (2012a). Late-life learning in Europe: Implications for social and public policy. In C. N. Phellas (Ed.), *Aging in European societies* (pp. 255–266). New York: Springer.

Formosa, M. (2012b). European Union policy on older adult learning: A critical commentary. *Journal of Aging and Social Policy, 24*(4), 384–399.

Formosa, M. (2014). Four decades of Universities of the Third Age: Past, present, and future. *Ageing & Society, 34*(1), 42–66.

Huang, C. (2006). The University of the Third Age in the United Kingdom: An interpretative and critical study. *Educational Gerontology, 35*(10), 825–842.

International Association of the Universities of the Third Age (AIUTA). (n.d.). *International Association of the Universities of the Third Age*. http://aiu3a.com/home.html. Accessed May 8, 2018.

Laslett, P. (1989). *A fresh map of life: The emergence of the third age*. London: Weidenfeld & Nicholson.

Manheimer, R. J., Snodgrass, D. D., & Moskow-McKenzie, D. (1995). *Older adult education: A guide to research, programs, and policies*. Westport, CT: Greenwood.

Midwinter, E. (1984). Universities of the Third Age: English version. In E. Midwinter (Ed.), *Mutual Aid Universities* (pp. 3–19). Kent: Croom Helm Ltd.

Midwinter, E. (2004). *500 beacons: The U3A story*. London: Third Age Trust.

Midwinter, E. (2005). How many people are there in the third age? *Ageing & Society, 25*(1), 9–18.

Philibert, M. (1984). Contemplating the Universities of the Third Age. In E. Midwinter (Ed.), *Mutual Aid Universities* (pp. 51–60). Kent: Croom Helm Ltd.

Picton, C., & Lidgard, C. (1997). Developing U3A. *Third Age Learning International Studies, 7*, 219–224.

Radcliffe, D. (1984). The international perspective of U3As. In E. Midwinter (Ed.), *Mutual Aid Universities* (pp. 61–71). Kent: Croom Helm Ltd.

Swindell, R. F., & Thompson, J. (1995). International perspectives on the U3A. *Educational Gerontology, 21*(5), 415–427.

Third Age Trust. (n.d.). *Third Age Trust.* http://www.u3a.org.uk/. Accessed May 8, 2018.

University of the Third Age Cambridge. (n.d.). *University of the Third Age Cambridge.* http://u3ac. org.uk/. Accessed May 8, 2018.

Vellas, P. (1997). Genesis and aims of the Universities of the Third Age. *European Network Bulletin, 1*, 9–12.

Withnall, A. (2016). United Kingdom. In B. Findsen & M. Formosa (Eds.), *International perspectives on older adult education: Research, policies, practices* (pp. 467–480). Cham, Switzerland: Springer.

François Vellas is the President of the International Association of Universities of Third Age (AIUTA) and Director of the University of Third Age of Toulouse (France). He is UN/EU/ WTO Expert in Tourism Policy and Strategy for the Aged. He is also Professor of International Economics at the University of Toulouse Capitole and holds a Ph.D. in economics. Each year, he chairs the WSTC World Senior Tourism Congress. He is the author of a dozen books and reports on older persons and tourism, a founder member of the UNESCO Chair UNITWIN, and was appointed by G20 secretariat to prepare the 'G20 Tourism Report'.

Part II
European Perspectives

Chapter 3
"An Alternative Ageing Experience": An Account and Assessment of the University of the Third Age in the United Kingdom

Keith Percy

Introduction

The University of the Third Age (U3A) was founded in the United Kingdom (UK) in 1981 and has grown steadily over the past thirty-seven years, currently reporting more than 400,000 members (U3A, 2018). The UK U3A was loosely inspired by the French *Université du Troisième Âge* but, as a founding fundamental principle, rejected close association with either an existing university or with national and local governments (Midwinter, 1984a). Other fundamental principles included the creation of self-determining local groups, cooperative peer learning for pleasure and interest, no boundaries to the range of topics and activities, no awards of credit or qualification and with members defined not by chronological age but by being no longer in full-time employment (ibid.). The U3A, despite its enormous growth, has not departed after thirty-seven years from these fundamental principles. Indeed, the headline slogan in its 2018 promotional material is "Learn—Laugh—and Live". This chapter traces the origins of U3As in the UK, with special reference to the British movement's ideological standpoints. The U3As' political climate, student profile, curriculum, and teaching and learning approach are all discussed, as is the organisation's future issues and challenges.

Origins and Development

In the late 1970s, there was a regular publication of official statistics showing a significant projected increase in the proportion of older people in the population of the UK. It caused considerable interest and disquiet among a number of educators,

K. Percy (✉)
Lancaster University, Lancaster, UK
e-mail: kozmapercy@gmail.com

© Springer Nature Switzerland AG 2019   33
M. Formosa (ed.), *The University of the Third Age and Active Ageing*, International Perspectives on Aging 23, https://doi.org/10.1007/978-3-030-21515-6_3

some of them in the adult education departments of various universities. They called for and suggested new ideas for the access of older people to learning opportunities, research into learning in later life and policy and funding from the government which would facilitate greater involvement in late-life learning. Their arguments did not rest solely on social justice and the entitlement of older people, but also on older people as an important resource for society.

Playing a singular part in this mix of reflection and action were three men who are still recognised, rightly, as the founders of the University of the Third Age in the UK (Midwinter, 2007). Michael Young was a distinguished social activist and politician, a socialist, who had a remarkable career as a social reformer. He founded, or assisted in the foundation of, a multiplicity of socially important organisations. Peter Laslett, a Cambridge University historical demographer and historian of political ideas, became an ardent practical reformer in later life. Both men were over 60 years in the late 1970s. Eric Midwinter was younger and had worked with Michael Young. Social historian and social policy analyst, community educator and social activist, Midwinter became a prolific author of books on education, history, sport and comedy. Together, these three identified the UK's ageing population as a key social issue and examined the French idea of the *Université du Troisième Âge* which had originated at Toulouse in 1973. Influenced by their collective experience of social reforms which developed from the grass roots, without the assistance of the state and established institutions, they concluded that a British variant of the *Université du Troisième Âge* relying on self-help, independence and mutual aid was desirable. Laslett, in particular, had negative experience of trying to persuade Cambridge University colleges to assist older people in their educational aspirations. However, a scheme was formulated for a first U3A group to be developed at Cambridge in 1981 and eventually launched in the autumn of 1982. This British variant of the *Université du Troisième Âge* would be open to all but with emphasis on those not in full-time employment or with a dependent family. It would be self-supporting financially, formulate its own curriculum, not award qualifications or pay teachers and encourage teacher-less peer learning. The Cambridge experiment began with the underlying aim that it would be the basis of national development.

"All our Futures"—the prospectus of the University of the Third Age in Cambridge—written in mid-1981, primarily by Laslett, was enormously ambitious and contained both objects and principles which would soon underpin the development of the U3A in the UK (Midwinter, 2004). One of its five "guiding principles" was that "the University of the Third Age shall consist of a body of persons who undertake to learn and to help others to learn. Those who teach will be encouraged also to learn and those who learn shall also teach" (ibid.: 738). One of its "Objects" was "to organise this institution that learning is pursued, skills acquired, research openings pursued and intellectual interest developed… without reference to qualifications, awards or personal advancement" (ibid. 741–2). After a national seminar in Cambridge in March 1982 and the achievement of a small grant from a national charity, there commenced the next stages of what became a classic national roll-out of a social innovation. A part-time national organiser (working from home) was appointed. She proceeded to develop a newsletter which would be distributed nationally but especially to those who had indicated an interest in starting some kind of informal local learning group or had already done so. A "start-up kit" was quickly developed centrally to assist such people and mentions were made in the national media, including a national radio slot to which 400 people responded in writing. Thereafter, events moved very quickly. The Cambridge U3A grew quickly with scores of learning groups and hundreds of members (each paying £20 for annual membership). In London, a U3A group also developed strongly with practical help from London University and a membership fee of only £4. In Huddersfield, in the north of England, a new U3A had "support from the local authority, hundreds of members and a weekly dance" (Beckett, 2009: 2). In twelve months, several other smaller local groups also emerged in rural areas. It is noteworthy that these times were not devoid of some public opposition to the U3A movement. In Lancashire, for example, adult education professionals opposed the emergence of a form of education for older adults which did not remunerate tutors and which did not try to protect standards of teaching (Percy, 1984: 152). Midwinter (2007: 4) even related a story of a meeting in Scotland in which he was "told in very choleric and outraged terms that it was…dangerous to permit lay people to organise their own educational affairs".

In October 1983, a decisive step was taken in that it was decided that if the U3A idea was going to spread nationally and new local groups mentored into existence, further grants or other sources of income would be necessary to finance promotional work, communication, travel and salaries. It was agreed that this scenario would require a structure with a legal status that charities and other funding sources could recognise. A U3A national organisation called *The Third Age Trust* was registered both as a charity and as a company limited by guarantee, and local groups started or planned under the U3A umbrella were invited to become members of the Trust. By the end of 1983, eight local groups had joined the Trust, which proceeded to request and secure grants from several charities. In the subsequent years, the U3A was beset with uncertainties and conflicts as it tried to find a viable balance between different goals and interest groups. Although it continued to grow in numbers of members and local groups, it was constantly troubled by the need to generate adequate finance, disagreements over the use of the label "university" and the appropriateness of links to established universities and colleges. It was certainly difficult to find a representational structure which would satisfy all local, regional and national interests, and at a conference of local groups in 1985, representatives were told that the Trust was almost insolvent. With some heart-searching, it was decided to suspend temporarily one of the founding principles and to seek short-term financial support from the government. The bid was successful and an unconditional grant of *circa* £15,000 over 3.5 years received to establish a national office, a newspaper and to develop new U3As.

Although further funding was needed, there was dissension among member groups over paying subscriptions to the Trust. In 1987, it was agreed that all U3A groups with more than 100 members should contribute £50, irrespective of size. This was obviously inequitable, and a *per capita* fee of £1 per local member was agreed at the next conference. However, this penalised larger groups with a significant increase and, in some cases, having to pay subscriptions of, say, £1000 (or, in the case of Huddersfield, nearly £2000) rather than £50. In 1988 and 1989, four groups disaffiliated (including two of the largest, Cambridge and Huddersfield). The split meant a loss to the national organisation of a third of its membership and a third of its potential income. If the London U3A group, which was large and a substantial financial contributor, had also left, the Third Age Trust might have imploded. U3A might then have continued as a loose assembly of local groups for a while but, without a central focus, would probably have disappeared. However, London remained and the Third Age Trust limped into the 1990s with about 100 local groups, about 8000 members and an uncertain future (Midwinter, 2004).

It is remarkable that, by the end of the 1990s, the U3A in the UK had recruited about 100,000 members and had largely assumed the shape and *modus operandi* which persist today. It had adopted a mission of national coverage by local groups which would be supported centrally while retaining their independence. It made real the paradox of professionalising itself while retaining an identity as a self-help and grass-roots organisation. There is no doubt that the social malaise of the 1990s in the UK had assisted this transformation. Political ineptitude, economic instability, social dissension and intellectual bankruptcy were fertile ground for the growth of an organisation such as the U3A which had developed a clear and self-reliant mission. As state-provided and state-financed adult education opportunities withered away, so the U3A flourished. In the twenty-five years or so since the early 1990s, growth and gradual increases in the member subscription levied on local groups have greatly increased the core income of the *Third Age Trust*. Moreover, the Trust achieved a number of financial coups, garnering significant fixed-term grants and finance from charities and foundations, the National Lottery, government and even a travel company, to continue expansion, develop services, professionalise administration and acquire adequate offices. There also began a steady emergence centrally of a range of educational services and learning support for local groups and members which were individually important but also, collectively, helped to define, explain and legitimise the role of and value of the Trust (see below under "Teaching and Learning"). When the mid-2000s witnessed a further increase in expenditure cuts in public educational opportunities for adults, older adults headed towards the U3A in even greater numbers. Since the mid-2000s, U3A membership has increased by 9–11% annually. Indeed, in 2004 the U3A in the UK had 539 local groups and 141,301 members; in 2008, there were 716 local groups and 209,000 members; in 2010, membership went past 250,000; and in 2018, there are over 1000 groups and 400,000 members.

Ideology Standpoints

Midwinter (2007) described Laslett, Young and himself as "Utopian or Ethical socialists". This is an important and noteworthy symbolisation as the ideologies underlying the British U3A movement were not purely or primarily educational but inherently socialist, communitarian and utopian—based on notions of self-help and mutual aid.

These notions had been put into practice before, in different settings, particularly by Young. They implied people doing things for themselves, cooperatively and independently of the state and the state apparatus of institutions dependent on the state. The Pre-School Playgroups Association was one example in which parents in the 1970s grouped together locally to organise childcare without waiting for local authorities to find the money and the will to meet an evident social need. Another example was the Consumers' Association in which consumers worked together to exercise and protect consumer rights against the giants of the marketplace and in the absence of state protection. Young's enemy, reflected Midwinter (2007: 3), was "giantism".

Midwinter, Young and Laslett agreed in the late 1970s that access to education by older people was the urgent social issue to which they should turn their attention. They thought that neither the state nor the "giants" responsible (i.e. the universities and the colleges as part of the state apparatus) were likely do anything for older people in this regard. If the notions of learning for qualifications and of professional and paid teachers were set aside, then groups of learners could be free and empowered to learn what, when, where and how they wanted. It was a set of ideals and implied procedures, they thought, perfectly suited to older people. However, the organisation was everything, with the local group being the key unit, independent and able to make its own decisions. The role of any network or uniting structure external to the group was to offer (but not insist on acceptance of) advice, opportunities, material support and perhaps learning materials. However, the U3A ideology proved to be made up of many ideas, some which did not necessarily cohere. One of the ideas was essentially missionary, that of ceaseless activity to create local groups in every part of the country and to collaborate with them (Laslett, 1989). For the U3A, that implied organisation, money and people doing it. It also meant raising money in the form of grants (including from the state) and subscriptions (from the members). Midwinter (2007: 3) always recognised, as U3A grew, the potential tensions inherent in its structure and the difficulties of pursuing side by side all of its original ideals. The *Third Age Trust*, of course, had to operate as a charity and as a company (for that is what it was). But, above that, was its overarching purpose of sustaining sovereign local groups through which older learners could learn, in their own way, what they wanted to learn.

Aspects and Characteristics

Political Climate

Over the years, UK governments largely ignored the education of older people. Ironically, in 2009, the government of the day published a progressive policy paper on informal adult education in which it mentioned that "with an ageing population, we know that learning for leisure in later life can maintain mental and physical health and contribute to people's sense of wellbeing" (Department for Innovation,

Universities and Skills, 2009: 7). One year later, the government fell from office and its promising ideas went with it. Nevertheless, over the recent decade or more, there have been a sequence of government-sponsored statements and reports on ageing, though not on education per se, but on the health and well-being of older people such as *Ageing well* (Department of Health, 2014) and *Living well in older years* (ibid., 2017). These policy reports have emphasised the importance of positive and active ageing and the significance of social and mental activity for older people. One report, the *Future of an ageing population* (Government Office for Science, 2016: 8), actually had a short chapter on the importance of lifelong learning for older people, citing a range of benefits such as improved "mental capital, which in turn increases resilience in later life...[and] can also help improve physical and mental health, reducing pressure on family and community resources". However, the authors, somewhat despondently, predicted that publicly financed learning opportunities for older people would decrease rather than increase in the future.

Student Profile

A mapping exercise of U3A members in 2001 showed their average age to be 70.6 years. As many as 6% were aged under 60, 39% aged in the 60–69 cohort, 42% aged in the 70–79 age cohort and 13% aged over 80. Women greatly outnumbered men, with 74% of members being women. While 65% of members reported no "recent learning activities" prior to joining U3A, 23% disclosed that they had no "formal educational qualifications". Apart from a splendid 4% of members who claimed attendance at U3A activities 10 or more times per month, the remainder split relatively evenly into thirds—one-third claiming attendance 5–9 times per month; one-third 3–4 times per month and the remaining third once or twice per month The UK U3A movement was shown to have a largely middle-class membership in terms of members' occupational background with 40% professional, 15% managerial, 15% administrative, 4% technical and 4% manual and 5% having had occasional or no employment. Moreover, 84% of members had access to some form of occupational or private pension (Midwinter, 2004: 4275–80). Midwinter makes a good case for the view that dismissing the U3A membership as primarily middle class is overly simplistic and is using shorthand for a complex social phenomenon. He pointed out, for example, that the U3A has "in fact, established local groups in "non-middle class enclaves" and that a proportion of the middle-class members of U3A are "risen" working class (ibid.: 41457, 4163). There is evidence (largely from websites and local reports) of local groups reaching out for membership from unrepresented districts in their areas, minority groups, disabled people and older people in residential settings but with varying success. Definitive data on this issue are not available but the received wisdom has remained that the U3A is largely a middle-class institution. As early as 1982, Laslett (as cited in Norton & Bourne, 1982: 1) stated that "it is inevitable that the U3A will be rather middle-class in composition". Twenty-five years later, Midwinter (2007: 4) reflected on how in the early days "the poor old

U3A became the victim of its own immediate success… constantly berated for not doing everything for everyone…for the working classes, the ethnic minorities, disabled older people, housebound older people. The people who benefit from the U3A are those who benefit from the U3A".

Curriculum

The U3A ideology insists that the curriculum of a local group can include any subject, activity or interest with which a group of members wish to engage. There is a page on the 2018 U3A website which offers links to guidance, notes, articles and more on a list of 62 "subjects". The list contains an amazing variety. It ranges from Archaeology to Mahjong to Wine Appreciation; from Ballroom Dancing to Puppetry to Theology; from Classical Greek and Painting to Cycling (social) and Sailing and Water Activity; and from Languages and History to Bridge and Walking (U3A, 2018). In 2005, the U3A researched the interests and activities offered by local groups. The most popular activity proved to be walking and rambling with music appreciation a close second. Among a long list of minority interests and activities were, for example, Northumbrian pipe making, inshore navigation and surveying the local town's best eating places (Beckett, 2009). Midwinter (2004) cited a 1990s survey of local groups in which the ten most popular subjects accounted for half of the learning groups. They included art, crafts, history, indoor games (Bridge, Scrabble), languages, literature, music, science, social studies (including discussions) and walking. Laslett (1996: 7) wrote correctly that "it is characteristic of the voluntary U3A that it takes the liberty to not to feel obliged to conform to academic standards set elsewhere. The members set their own standards just as they choose the subjects of study and include many things not usually recognised as 'properly' academic".

Teaching and Learning

Midwinter (2007: 4) described the three founders as letting "the genie of self-mobilised learning out of the bottle". Although self-mobilised learning became a reality in the practice of local U3A groups, there are not many studies of this practice upon which to draw. Three available studies are, however, of interest. Marsden (2011) addressed the question of the "cooperative learning model", as she called it, of a U3A group in a northern English town. She used a particular methodology called phenomenography in which analysis of interviews with members describing their U3A learning experience allows the formation of a "pool of meaning". From the "pool", a structured set of categories could be derived to describe the different ways of experiencing the phenomenon of learning in the group. Marsden identified a hierarchy of four conceptions of cooperative or participative learning, articulated by the members, which she considered distinct and which formed a range from the least to the most participative—namely learning from an expert in the field; learning through didactic teaching from a keen amateur; learning through a variety of teaching modes; and learning as equal partners. Cheng (2013) conducted an observational and participative study of teaching and learning in an urban U3A in the English Midlands. She described the music group in which "members just sat comfortably enjoying music, piece after piece, occasionally they had a little discussion with the group leader" (ibid.: 22). In a history group, the group leader professed that "members can take in history through a lecture, but they can also learn and share something by listening to others' personal experiences" and that "the content they are learning is quite random" (ibid.). In a literature group, "the work…was actually a key to their memories that were awoken by those words and scenes, then they started talking about the most popular songs and people's dressing style, etc. in that particular time with great relish" (ibid.). The Latin class was a mixed-level group: "when the group leader was mentoring one group, members of other groups would study on their own, some members were just chatting away" (ibid.: 26). Cheng (ibid.: 23) commented that for these older learners "the quantity and the outcomes of their learning [are] not very important anymore, being in a physically and psychologically comfortable environment to 'enjoy' the process of learning becomes more important for them". On the other hand, a third study, an action research project in the North of England, showed that a sizeable proportion of members of one U3A local group felt that they could benefit from the challenge of academic study at a relatively high level. Over 7 months, members were offered the chance to participate in a range of learning opportunities on their local university campus. Some 149 members of this U3A group (which had about 750 members) joined the project, with 70% being aged 60–74 years and 56 of them sitting alongside undergraduates aged primarily 18–21 years in regular university lectures. As the project ended, asked by researchers to evaluate their experience, many of the U3A learners described its fascination and its challenge. It was intellectually exciting to be confronted by university research. The learning context, with its cosmopolitan and international nature, was a revelation. Some of the University teachers told the researchers that they valued the presence of the U3A learners because they had the confidence to ask and answer questions. When asked which of U3A or University teaching methods and contexts suited them better, most U3A members replied that the methods were complementary. It was good to go from one to the other and back again. Anyway, the reality was that there was a heterogeneity of teaching and learning approaches and contexts in both the University and the U3A (Percy, 2012). Indeed, that is probably the best way to consider the teaching and learning approaches and contexts in the U3A. There is a wide range and the decision on what takes place emerges in the local groups. That decision is supported by considerable and impressive learning resources, advice and opportunities available to members and groups from the national organisation. On-line there are the Resource Centre (an extensive DVD library service of educational resources available for use in local groups), Sources On-line (the subject-based educational journal of the U3A) and the National Resource Database (UK-wide research database). There are also national programmes of residential subject summer schools and the regional programmes of study days; a Travel Club of educational tours; National Subject Networks which link members with common interests and arrange seminar days and workshops; and the Shared Learning Projects (SLPs) in which research teams of members drawn from at least two U3A local groups work with an external organisation such as a library or gallery (U3A, 2018).

Finance

As of the year 2018, members of a U3A local group pay a modest annual fee in the region of £15, with a percentage of that fee forwarded to the Third Age Trust. When the member attends a learning group there can be a small additional fee to cover such costs as accommodation and materials, if relevant. However, many groups meet in private homes and there are no tutor fees. Attendance at a national summer school or a regional study day and subscriptions to Third Age Trust publications and other items are paid for separately. During the 2016–17 financial year, the Third Age Trust had an income of about £2.5 million. Income from the share of member subscriptions was about £1.35 million. Remaining income of over £1 million came primarily from the sale of *Third Age Matters*, the U3A newsletter and advertising. An examination of the Third Age Trust's Expenditure Account for 2016-17 shows that total expenditure was almost £2.5 million and that 40% of expenditure made centrally was on what was described as "supporting the learning of members and promoting the benefits of self-help learning". In addition, 19% was attributed to "providing advice and support for management in U3As" and 7% to "governance" (Third Age Trust, 2017).

Future Issues and Challenges

Two things stand out about the British U3A. It has enjoyed significant and increasing growth to date and, despite that growth, the structure, shape and processes which had been stabilised by the late 1990s have not changed significantly. It obviously now has at its disposal a successful model for continuing growth. However, social institutions rarely stay unchanged. It seems possible that sooner or later there could be attempts to translate U3A's ethic of grass-roots self-help and independence into other directions. Given the cost of publicly financed education, it may be that future politicians and civil servants will explore the idea of supporting replication of U3A contexts and approaches in other settings and for different purposes. Moreover, it was no accident that Young set up the Third Age Trust in such a way that it could provide a legal entity for social enterprises for older people, other than education, such as housing, communication, health, finance, design and architecture—basically, almost any aspect of social life (Midwinter, 2007). The principle would be of older people acting together, not waiting for action. It was, at least initially, in the minds of the founding trio that one or more of these other areas could conceivably in the future draw the U3A into wider action.

There is little doubt that the U3A in the UK benefits the physical and mental well-being of older people who participate as members of local groups—although the degree of that benefit will vary according to the nature of U3A participation and the individual's lifestyle (Swindell & Thompson, 1995). One can conclude confidently that participation in the U3A contributes significantly to positive and active ageing. Of course, these statements are general ones and the concepts need to be closely and operationally defined, tested and assessed. Nevertheless, both logic and an abundance of common sense and subjective evidence can be assembled to support the statements. Participation in U3A activities takes older people out of their homes, on a regular basis, to join with others in physical, cognitive, expressive, imaginative, explorative, altruistic, interactive and social activities. And, currently, there are 400,000 of such older people. In the newsletters of local groups and in the columns of such publications as *Third Age Matters* and *Sources,* there is a multitude of testimony to this effect. In June 2018, the U3A published a self-assessment which is significant for the way in which defines its own success. Noting that "there has been no recent evaluation of the benefits of the model of lifelong learning offered by the U3A", the publication offers reports on a literature review, a questionnaire completed by 801 participants and eight focus groups (Third Age Trust, 2018: 5). Its conclusion is that this research "demonstrates the value of mutual aid and of reciprocity" and that "the U3A model offers an alternative ageing experience, which is built on shared learning, skill sharing and volunteering" (ibid.: 19).

In conclusion, there is no doubt that "An alternative ageing experience" is a fine description. The U3A has, indeed, now become an important social institution in the UK while remaining true to its basic principles which have been sustained by pragmatic but logical strategies for development. However, at some point, it will have to confront the issues relating to the greater participation of older people from working-class and ethnic minority backgrounds in its activities. It is clear that in 2018 the University of the Third Age in the UK is still in development. It has achieved great success, but internal needs and external opportunities will arise—the challenges of which will need to be addressed.

References

Beckett, F. (2009). *The U3A story*. Bromley: The Third Age Trust.

Cheng, Y. (2013). *Learning in the Third Age in the UK: Motivators and barriers.* (Unpublished Master's thesis). University of Wolverhampton, Wolverhampton, United Kingdom.

Department of Health. (2014). *Ageing well. Compendium of fact sheets.* https://assets.publishing. service.gov.uk/government/uploads/system/uploads/attachment_data/file/277584/Ageing_Well. pdf. Accessed April 26, 2018.

Department of Health. (2017). *Living well in older years. Guidance.* https://www.gov.uk/government/publications/better-mental-health-jsna-toolkit/7-living-well-in-older-years.pdf. Accessed April 26, 2018.

Department for Innovation, Universities and Skills. (2009). *The learning revolution*. London: Department for Innovation, Universities and Skills, HMSO.

Government Office for Science. (2016). The *future of an ageing population*. Research and analysis. https://www.gov.uk/government/collections/future-of-ageing.pdf. Accessed April 26, 2018.

Laslett, P. (1989). *A fresh map of life: The emergence of the third age*. London: Macmillan Press.

Laslett, P. (1996). The intellectual independence of the Third Age. In J. G. Agius (Ed.), *Education for the elderly: A right or obligation?* (pp. 1–10). Msida: Malta University Press.

Marsden, R. (2011). A study of the cooperative learning model used by the University of the Third Age in the United Kingdom. *International Journal of Education and Ageing, 2*(1), 55–66.

Midwinter, E. (Ed.). (1984a). *Mutual Aid Universities*. London: Croom Helm.

Midwinter, E. (1984b). Universities of the Third Age: English version. In E. Midwinter (Ed.), *Mutual Aid Universities* (pp. 3–19). London: Croom Helm.

Midwinter, E. (2004). *500 Beacons: The U3A story* [Kindle DX version]. Retrieved from Amazon.com.

Midwinter, E. (2007). *U3Aology: The thinking behind the U3A in the UK*. Frank Glendenning Memorial Lecture, March 23, 2007. Leicester: NIACE.

Norton, D., & Bourne, R. (1982). *University of the Third Age: The National Conference*, March 26–27, 1982. Cambridge: Unpublished report.

Percy, K. (1984). The University of the Third Age in Lancaster, Morecambe and District. In E. Midwinter (Ed.), *Mutual Aid Universities* (pp. 146–155). London: Croom Helm.

Percy, K. (2012). Learning in later life: Universities, teaching, intergenerational learning and community cohesion. In *3rd Conference of the ESREA network on Education and Learning of Older Adults, Intergenerational Solidarity and Older Adults' Education in Community*, September, 19–21, 2012, University of Ljubjana, Slovenia.

Swindell, R., & Thompson, J. (1995). *An international perspective of the University of the Third Age*. http://worldu3a.org/resources/u3a-worldwide.htm. Accessed November 30, 2017.

Third Age Trust. (2017). *Annual report and accounts*. London: The Third Age Trust.

Third Age Trust. (2018). *Learning not lonely. Living life. Expanding horizons. Challenging conventions*. London: The Third Age Trust. https://indd.adobe.com/view/c99ad603-0622-4636-b12c-9f8ad363dae8. Accessed October 17, 2018.

University of the Third Age. (2018). *Third Age Trust/University of the Third Age*. https://www.u3a.org.uk. Accessed March 2, 2018.

Keith Percy is Professor Emeritus of Continuing Education at Lancaster University, UK, where he was Head of the Department of Continuing Education and Director of the School of Lifelong Learning and Widening Participation. He is currently Chair of the Association for Education and Ageing and Editor-in-Chief of the *International Journal of Education and Ageing*. A sociologist and historian, he has published widely over fifty years on adult and continuing education; history, policy and student progress in higher education; widening participation in higher education; educational gerontology; access to learning for older people; and well-being in later life.

Chapter 4
Be Active Through Lifelong Learning! The University of the Third Age in Iceland

Hans Kristján Guðmundsson

Introduction

The first University of the Third Age (U3A) in Iceland—U3A Reykjavik—was established in 2012. Its launch changed the concept of third age in Icelandic society into that of being a period of freedom, independence, activity, and enjoyment in the second half of the persons' life course. The concept of the 'third age' was not, until then, very well-recognised in Iceland. Previously, concepts and terms normally associated with ageing and later life—such as retirement, dependency, and long-term care—dominated discussions on ageing creating an idea of a group of citizens that constitutes a burden to the so-called productive generations. This chapter presents the background and social environment around the emergence and successful development of the first U3As in Icelandic society, with special emphasis on issues relating to those generations approaching or already living in their third age.

Demographic Trends

The age distribution of the Icelandic population is rapidly changing, and its demographic developments are highly similar to other international populations where the percentage of persons aged 65-plus exceeded the percentage of children aged less than 5 years (United Nations, 2015). Whilst life expectancy is increasing—with people living considerably longer than they did in the past—the number of persons in the younger generations is growing considerably slower. In 2016, the life expectancy in Iceland was estimated to be 80.7 years for men and 83.7 years for women (Statistics Iceland, 2017). In the same year, the life expectancy for those who have reached the

H. K. Guðmundsson (✉)
U3A, Reykjavík, Reykjavík, Iceland
e-mail: hanskr@simnet.is

© Springer Nature Switzerland AG 2019

M. Formosa (ed.), *The University of the Third Age and Active Ageing*, International Perspectives on Aging 23, https://doi.org/10.1007/978-3-030-21515-6_4

age of 65 reached 21.3 years for women and 19.5 years for men (Organisation for Economic Co-operation and Development, 2018a). Icelandic citizens can, on average, expect to enjoy a period of freedom and activity in retirement of a similar, or even of a longer, duration to adulthood.

In January 2018, the population of Iceland reached 348,450 citizens—growing by 3% from the preceding year (Statistics Iceland, 2018a). Whilst the number of Icelandic citizens aged 50-plus amounted to around 112,000 individuals, around 32% or one-third of the total population, the number of citizens aged 65-plus constituted close to 49,000 individuals, or 14% of the total population (Statistics Iceland, 2018b). According to the medium variant population projections issued by Statistics Iceland (ibid.), the Icelandic population is projected to increase by 28% so as to reach 447,236 by the year 2060. In comparison, the number of people aged over 50 is projected to grow by 76%, from 32 to 44% of the total population. At the same time, the number of people aged 65-plus is projected to grow by 140%, from 14% of the total population to 26% in the 2018–2060 period, whilst over the same period the 20–64 age group will increase by only 14% (ibid.). In national economic considerations, the old-age dependency ratio is normally used as a measure of the ability of working individuals, to provide for the finance of retirees. To this end, the working population has been defined as those within the age ranging from 20 to 64 years, and the 'old age dependent population' has been defined as those aged 65 years and older. The ratio is calculated as the percentage of the 'old-age dependent population' relative to the working population. Presently, this ratio in Iceland is approximately 25%, but it is projected to rise to 50% in 2060 (ibid.). This means that in the year 2060 for every 100 individuals within the 'working age' range there would be 50 individuals aged 65 years and older. The 'old-age dependency ratio' is frequently used to estimate the 'burden' that the workforce is supposed to carry on its shoulders, in order to provide for a steadily increasing number of older citizens.

Labour Market Participation and Income Security

Labour market participation by older people in Iceland is relatively high compared to that of other countries. Statistics show that in 2016 the labour force cohort (16–74 years) reached 196,500 workers—that is, an 83.6% activity rate (Organisation for Economic Co-operation and Development, 2018b). With respect to the cohort of persons aged 55–74 years, the activity rate was 67.7%, whilst for people aged 65-plus this figure stood about 40% which is by far the highest in the world. In Europe, Estonia comes closest at 26%, followed by Norway at 19%. The OECD average was estimated to be that of 14.5%. Looking at gender differences, activity rates for women were lower than those for men, with rates decreasing more rapidly as age increases. It ensues that when compared to other countries, the Icelandic population seems to stay in the labour force for longer. These statistics are also in harmony with the estimated average age of retirement in Iceland which in 2016 was approximately 70 years for men and 67 for women (ibid., 2018c).

The required statutory age to receive a public pension is 67 years, although the law allows for some flexibility. Having said that, employees in public service cannot hold their positions beyond the age of 70. The Icelandic pension system is based on three pillars. The first pillar is a tax-financed public pension scheme, the second pillar consists of mandatory savings-based occupational pension schemes (pension funds) operated by the labour market partners, and the third pillar incorporates voluntary private pension savings with tax incentives. The occupational pension system is mandatory and contribution-based and has been operating since the beginning of the 1970 s. The third pillar has been operating only since the late 1990s. The interplay between the first and second pillar was radically changed by new legislation that entered into force in January 2017, where the public pension scheme was simplified by unifying three pension benefits into one old-age pension with some controversial income-linked rules (Olafsson, 2016). Changes included the abolishing of a previous non-income related basic pension benefit, allowing for the income/benefits from the occupational funds to reduce the public benefits. In practice, as a result, the occupational pension scheme has been defined as being the first pillar, and the public scheme to be secondary. This new legislation also increased the public statutory retirement age from the present age of 67 to 70, for both men and women, over a period of 24 years. However, the new regulations allow for greater flexibility. On the one hand, it will be possible for Icelandic citizens to retire and receive their public pension at age 65 instead of 67, albeit with reduced pension entitlement. An incentive for delaying the receipt of one's pension to age 80 is now also available, with the inducement being a yearly increase in pension entitlement. Such policy changes have—as a result—changed the available options when one is planning one's exit from the labour market. Whilst on the one hand incentives to continue working were significantly reduced, on the other hand more flexibility has been introduced in as far as the age of retirement of Icelandic citizens.

Education

The Icelandic education system includes a non-compulsory playschool period, followed by compulsory basic education (primary and lower secondary) from 6 to 16 years, and an upper secondary stage from 16 to 19/20 years (matriculation, vocational, or equivalent). The tertiary educational stage encompasses university studies. In 2016, educational attainment amongst Icelanders aged between 55 and 74 years was as follows: 33% had attained primary education; 38% had attained secondary education; and 28% succeeded in enrolling in tertiary studies. With respect to citizens aged 65–74 years, the corresponding percentages were 45, 36, and 19%, respectively (Statistics Iceland, 2018c). In previous years, public schools offered middle-aged and older persons the possibility to enlist in secondary education, referred to as 'seniors' departments'. This was very popular amongst people who had already spent many years in the labour market but possessed no qualifications. Several people in their third age did benefit from this option and continued pursuing an academic career. A few years ago, however, a new legislation on adult education was passed, making it difficult for secondary schools to offer such opportunities, forcing potential third age students to seek out private institutions. At the time of writing, although several learning possibilities are available through programmes offered by trade unions, other non-governmental associations, and private institutions seeking to bridge the gap towards secondary and tertiary levels of education, it remains that there is no specific state policy regarding the possibility of adult education for older citizens or third age groups. On a positive note, the tertiary education system—comprising all the universities in Iceland—does not discriminate on age. If one has the required entry qualifications to enter a university, one can commence his/her studies at whatever age, given that the possible pre-knowledge requirements are met. Hence, it is common for people in their sixties, and seventies to enrol for masters and doctoral programmes and graduate with success. The Icelandic government also runs a Student Loan Fund, which is the primary funding institution for students, and which does not discriminate by age, although many Icelandic people do not think that it is a good idea to incur debts in your third age.

Government statistics on lifelong learning—gathered through labour force surveys—reported that in 2016, about 10.6% of persons aged between 55 and 74 years (67,600 persons) engaged in one or more learning programme outside the formal educational system, whilst 1.1% had participated in one or more learning programmes in schools (unfortunately, a downward trend in the lifelong learning attendance in the previous years) (Statistics Iceland, 2018d). Here, it is noteworthy that several funding opportunities are offered for lifelong learning by trade union and employee associations, although these funds are normally available to paying members only, and such opportunities are not available after retirement. Other options are available from the Icelandic Confederation of University Graduates (BHM) who run a career development centre that offers financial support to members to develop their career through further educational programmes. Many members, aware of impending retirement, utilise such an option to further their education in preparation for their third age. The establishment of the U3A in Reykjavik was based on the assumption that such an organisation—based on mutual sharing of knowledge and experience—would secure a prominent role in this rather diverse educational environment for adults and persons in the third age, and thus serve as a valuable and much-needed forum by complementing other opportunities for lifelong learning.

The Third Age

The third age concept is generally considered to be the age after retirement—sometimes also referred to as the golden years of adulthood—more or less spanning from age 65 to 80 (Laslett, 1989; Barnes, 2011). This period has been described as a period of adulthood with typically fewer responsibilities, adequate financial resources, and good physical health, offering rich possibilities for self-fulfilment and purposeful engagement, possibly lasting a few decades before years of biological and functional decline set in (Formosa, 2010, 2016). This young-old or third age is relatively new in human history and is closely linked to the increased life expectancy and the demographic developments in our societies. In Findsen and Formosa's words, the third age

> ...was first born and conceived in the 1950s to counteract the stereotype of later life as a short period which is plagued by illness, invalidity, and in most cases, poverty...The 'third age' refers to a specific socio-demographic trend within population ageing. It alludes to how the combination of increased longevity and a number of other social factors - ranging from earlier retirement, improving health status, establishment of the welfare institutions of retirement and pensions schemes, to more positive values and beliefs towards older persons - have opened up what could be loosely termed as a new phase in life, in which significant numbers of older persons spend a considerable amount of time in relative active years following exit from work.
>
> (Findsen & Formosa, 2011: 48–9)

The age ranges mentioned should be considered to be dynamic and fluid, and individually biased. The understanding of the term third age differs between countries, having both positive and negative connotations, and more often than not is linked to retirement age. The position adopted by U3A Reykjavik is not to use any fixed age interval to define the third age, as members are aware that the real age of retirement and the age of exit from the labour market can differ markedly between individuals, as well as from the statistical definition of the dependency ratio. Hence, the beginning of the third age is considered to be highly varied and to depend on the unique life course of each individual. The U3A Reykjavik targets all persons aged 50-plus, whilst its activities are open to anyone irrespective of age. After all, the fifties comprise the age when the life situation for most individuals starts to change, when many begin to think ahead, focusing especially on the years left in the labour market, or for the more adventurous, the years left before a career change becomes impossible. We should not forget that the experience and knowledge of this group of people are important for the progress of society.

U3A Reykjavik

Foundation

U3A Reykjavik was established on 16 March 2012, and thus, has the accolade of being the first third age university in Iceland. The founder of U3A Reykjavik, and its first chairperson, was Ingibjorg R. Guðjónsdóttir, who discovered the U3A movement on the Internet and then attended a World U3A conference in India in 2010. On her return, she established contacts that gave her excellent guidance towards developing the overall aim, objectives, purposes, and logistical requirements of the first U3A in Iceland. Two other influential persons in this foundation process included Ian Funnell from the UK's Third Age Trust, and Tom Holloway who was highly active in the

Indian U3A movement and active in the World U3A. Although the U3A concept is still relatively new in Iceland, in recent years it has been extremely well received by older Icelanders and Icelandic society in general. The reputation and visibility of U3A Reykjavik are steadily increasing and are viewed as an important forum for people in the post-50 years of age who want to stay active in the latter part of their lives. U3A Reykjavik is modelled on the anglophone model, as an independent association of people in their third age, regardless of their educational or professional background, and operates in the spirit of Laslett's original objectives for British U3As:

> …to educate British society at large in the facts of its present age constitution and of its permanent situation in respect of ageing…to create an institution for those purposes where there is no distinction between those who teach and those who learn, where as much as possible of the activity is voluntary, freely offered by members of the University…to undertake research on the process of ageing in society, and especially on the position of the elderly in Britain…(Laslett 1981, quoted in Formosa, 2014: 45)

Its aim is to provide its members with a diversified selection of learning activities—thus, serving as a platform for the learning and mutual dissemination of knowledge as an end-in-itself. The learning programmes are mainly based on peer-learning activities, and there are no educational requirements for enrolment and no examinations are taken. This classifies the U3A in Reykjavik as non-formal adult education, a lifelong learning activity aimed at people in their second half of the life course. In fact, at no point does the Reykjavik U3A use the term students but always refers to the attendees as members, especially as there are no formal links to any university, apart from an informal agreement with the Reykjavik University to access their Open University programmes. The U3A in Reykjavik is also a member of the Reykjavik Academy which is an association of independent scholars.

Growth and Activity Programme

In the 2012–2018 period, U3A Reykjavik experienced a steady increase in membership, with numbers increasing from 31 at the first founding meeting to 573 in September 2018. The membership fee has always been kept at a nominal level, whilst a low entrance fee is charged at most organised events. Women constitute 78% of members, and the average age is 69 years. More than half of the members are 70 years and younger, and as much as 66% are between 65 and 75 years of age. The mean age distribution of members is similar for both men and women. The information on U3A activities is generally spread through an informal yet wide-reaching grapevine, spearheaded by facilitators and lecturers who take on learning programmes on a voluntary basis. The U3A in Reykjavik has a highly active and visited website—www.u3a.is—which is also the main source of information on learning programmes taking place in the upcoming months. This website also enables viewers to register for an event, read news on U3A activities, and learn more about the ethos of the U3A movement. At the same time, the website disseminates information on international activities concerning the U3A movement, research projects related to the third age and older adult learning, and activities organised by the World U3A and the International Association of the Universities of the Third Age (AIUTA) (n.d.). Events, normally open to all, are also announced on the Facebook page and through direct electronic mailing to the members. U3A Reykjavik organises four different forms of activities:

- Weekly Tuesday events during wintertime. These take the form of lectures on widely diverse subjects, such as science, biology, innovation, social and political issues, environment, history, literature, and other cultural issues. The events are normally both educative and entertaining, and the lecturers are experts in their fields, often university teachers and members of the U3A Reykjavik.
- Monthly discussions in cafés or similar venues. The chats are informal meetings, where a keynote speaker leads the chat and everyone is free to participate in the discussion over a cup of coffee or tea.
- Guided visits and lectures to museums, educational spaces, such as historic or archaeological sites, and locations related to art and literature.
- Shorter courses that run through the whole semester on writing memoirs, pre-retirement learning and organisation, and how to make the best use of social media.

The members are encouraged to form interest groups on chosen subjects. In 2018, such interest groups included ones on literature, cultural visits, and international activities. Indeed, the U3A in Reykjavik is an umbrella for cooperative study and research projects on interesting subjects, and the successfully completed *European BALL* project and the *Catch the BALL* project—described below—are good examples of research teams under the auspices of the U3A in Reykjavik.

The second Icelandic U3A organisation, U3A Sudurnes, was founded on 16 September 2017, this time in Suðurnes, the South West region of Iceland. U3A Suðurnes now counts 50–60 members and includes three subject groups, meeting regularly during the winter months to discuss and/or listen to lectures on topics such as the geography of Suðurnes, its fauna and flora, culture and history, genealogy, and any other subjects that members may be interested in. There are many other subject groups in the pipeline, especially on poetry and literature. U3A Reykjavik has heartily supported the formation of this new U3A, and one hopefully anticipates the formation of further U3As in other regions of Iceland.

International Cooperation

U3A Reykjavik is one of the recent and smallest members of the AIUTA (AIUTA, 2018). This membership has assisted U3A Reykjavik to establish links to U3As in other continents and countries, resulting in many positive exchanges and bilateral contacts, as well as participation in AIUTA meetings and conferences. U3A Reykjavik has also developed international contacts through other associations. One of the most important is the participation in the Pass it on Network (2018), since such a network involves the exchange of ideas from all over the world on how to promote and support active ageing. These international networks have been crucial to engendering an Icelandic U3A temperament, enriching the opportunities available for healthy and active learning for third agers. Most importantly, such networking was crucial in devising learning programmes that aid the transitioning of middle-aged persons into the third age by enabling the carrying out of the *Be Active through Lifelong Learning (BALL)* and *Catch the BALL* projects with support from the European Union's Erasmus+ programme.

The 'BALL' Story

The BALL project (*Be Active through Lifelong Learning*) aimed to develop guidelines and recommendations on early preparations for a dynamic third age. The idea was first formulated in 2013 as part of discussions on the ethos and future strategy for U3A Reykjavík. The project was accepted for funding by the European Union's Erasmus+ programme and commenced in September 2014 involving three U3A organisations—namely, U3A Reykavik, the UPUA,Universidad Permanente of the University of Alicante (Spain), and the University of the Third age in Lublin (Poland). The project results, guidelines and recommendations were published in the book *Towards a Dynamic Third Age* in September 2016 (Guðjónsdóttir et al., 2016). The book and a few interim research reports are available and can be downloaded from the project website (Be Active through Lifelong Learning, 2018). As it was concluded,

> The results of these studies and exercises in the partner countries were found to be amazingly similar, given the differences between the societies. Similar attitudes towards the third age and retirement were seen, showing overwhelmingly positive expectations for an active and enjoyable life. It was obvious that existing preparation activities for retirement were similar in all the countries, mainly consisting of practical advice shortly before retirement. It was found that almost all respondents of the surveys considered preparations to be necessary and that they mainly wanted to be in charge themselves. Very few, however, said they did or would prepare. These similarities made the comparative studies easier to conclude and the guidelines proposed here can be seen as very adaptable throughout Europe.
>
> (Guðjónsdóttir et al., 2016: 69)

The BALL recommendations are defined in three actions, a three-step process proposed as a 'best practice' system for preparing individuals for the changes, challenges, and opportunities lying ahead in the third age. These were:

- *Awareness Raising* on the value and importance of the third age directed both towards society and the individuals themselves.
- *Personal Development Academy* which supports individuals who need and want to evaluate their strengths, desires, and possibilities to be prepared for the latter part of their lives.
- *Warehouse of Opportunities* for the Third Age which consists of a virtual online portal, where individuals get access to relevant opportunities.

The BALL guidelines have been disseminated internationally and have been very well received. Indeed, the project received the Erasmus+ quality award in the field of adult education in November 2016. This general appreciation has given a large boost to the work and recognition of the U3A Reykjavik both in Iceland and internationally. Research and recommendations on the benefit of the third age will consequently be high on the future agenda of U3A Reykjavik. The logical step forward was therefore to embark on a new project implementing these recommendations.

The 'Catch the BALL' Project

The aim and objectives of the *Catch the BALL* project were defined in 2016, in order to implement the BALL guidelines and recommendations. The project also received continued support from the Erasmus+ programme and was launched in December 2016. New partners joined the Icelandic BALL partners, the Kaunas Science and Technology Park (Lithuania), and the MBM Centre, Training and Development in Liverpool (UK). The Catch the BALL project (Catch the BALL, 2018) was finalised in July 2018, and the two main outputs were launched and introduced at its final conference in Reykjavik on the 26th of June. U3A Reykjavik was in charge of the development of the *Warehouse of Opportunities*, which is a web portal in the image of a real warehouse with web pages as racks and shelves, presenting opportunities as products on shelves. The aim is to offer people aged 50 and over a user-friendly gateway to diverse opportunities to improve their skills and life fulfilment, get advice on finances, on starting a new career, and on other issues pertinent to a dynamic third age. The European Warehouse of Opportunities is now open and running under the direction of U3A Reykjavik and can be accessed at its web link (Warehouse of Opportunities, 2018). Three linked national warehouses, the Icelandic, the Lithuanian, and the British, can be accessed through the European Warehouse. We hope to see more national warehouses to be established in the future. Another output of the project is a manual for trainers and facilitators, the *Academy of Opportunities*, giving ideas to those who would assist people in later life with assessing their possibilities for a prosperous and enjoyable future. This manual can be accessed through the warehouse (Academy of Opportunities, 2018).

Future Perspectives

There is no doubt that the U3A concept has arrived to stay in Icelandic society, as the organisation and its activities are steadily gathering momentum with each coming month. The message directed to the 50+ generation—that it is never too late to rethink your situation in life, embark on new things, fulfil your ambitions, and act upon your dreams and desires—has been enthusiastically received by all of Icelandic society irrespective of age. As the number of people in their third age will increase faster than the younger generations, it is important that society does not perceive the older generations as a burden, but celebrates their potential for creativity and productivity. The third age generation of the future will continue to live longer, be healthier, and will constitute a valuable human resource of knowledge and experiences strongly needed by society. U3A Reykjavik will meet future challenges to promote positive ageing in Iceland and worldwide, armed with experience and ideas after seven years of operation. The messages and recommendations developed in the BALL's and Catch the BALL's research together with the Warehouse of Opportunities will surely guide us towards a prosperous life in the third age.

References

Academy of Opportunities. (2018). *Manual for trainers and facilitators.* https://xn--vruhs-tkifranna-3lbd1zxc.is/wp-content/uploads/2018/09/Academy-of-Opportunities-Manual-for-Tra iners-and-Facilitators.pdf. Accessed September 14, 2018.

Barnes, S. F. (2011). *Third age: The golden years of adulthood.* http://calbooming.sdsu.edu/documents/TheThirdAge.pdf. Accessed May 24, 2018.

Be Active through Lifelong Learning. (2018). *BALL: Be active through lifelong learning.* https://www.ball-project.eu/. Accessed April 30, 2018.

Catch the BALL. (2018). *Create a dynamic third age.* http://catchtheball.eu/. Accessed September 14, 2018.

Findsen, B., & Formosa, M. (2011). *Lifelong learning in later life: A handbook on older adult learning.* Rotterdam, Netherlands: Sense Publishers.

Formosa, M. (2010). Universities of the Third Age: A rationale for transformative education in later life. *Journal of Transformative Education, 8*(3), 197–219.

Formosa, M. (2014). Four decades of Universities of the Third Age: Past, present, and future. *Ageing & Society, 34*(1), 42–66.

Formosa, M. (2016). Malta. In B. Findsen & M. Formosa (Eds.), *International perspectives on older adult education: Research, policies, practices* (pp. 261–272). Cham, Switzerland: Springer.

Guðjónsdóttir, A. M., Guðlaugsdóttir, I. R., Guðmundsson, H. K., Bru Ronda, C., Aleson-Carbonell, M., & Stanowska, M. (Eds.). (2016). *Towards a dynamic third age: Guidelines and recommendations of the Erasmus+ BALL Project.* Poland: Towarzystwo Wolnej Wszechnicy Polskiej Oddzial w Lublinie. https://www.ball-project.eu/sites/default/files/BALL_PublKonc_Internet/Ball_en2_023_200.pdf. Accessed May 25, 2018.

International Association of Universities of the Third Age. (n.d.). *International Association of Universities of the Third Age.* http://www.aiu3a.com/. Accessed April 22, 2018.

Laslett, P. (1989). *A fresh map of life: The emergence of the third age.* London: Macmillan Press.

Olafsson, S. (2016). *Iceland: New legislation reforming the public old age pension* system. ESPN Flash report 2016/67. http://ec.europa.eu/social/BlobServlet?docId=16456&langId=en. Accessed April 29, 2018.

Organisation for Economic Co-operation and Development. (2018a). *Life expectancy at 65.* https://data.oecd.org/healthstat/life-expectancy-at-65.htm#indicator-chart. Accessed April 29, 2018.

Organisation for Economic Co-operation and Development. (2018b). *Labour force participation rate.* https://data.oecd.org/emp/labour-force-participation-rate.htm. Accessed April 29, 2018.

Organisation for Economic Co-operation and Development. (2018c). *Aging and employment policies—Statistics on average effective age of retirement.* http://www.oecd.org/els/emp/average-effective-age-of-retirement.htm. Accessed April 29, 2018.

Pass it on Network. (2018). *A global network for positive ageing.* http://passitonnetwork.org/ Accessed April 22, 2018.

Statistics Iceland. (2017). *Life expectancy in Iceland is one of the highest in Europe*. https://www. statice.is/publications/news-archive/population/life-expectancy-and-mortality-rates-2016/. Accessed April 22, 2018.

Statistics Iceland. (2018a). *Icelandic population grew by 3% last year*. https://www.statice. is/publications/news-archive/population/population-1st-of-january-2018/. Accessed April 29, 2018.

Statistics Iceland. (2018b). The numbers ant percentages given are calculated by the author based on the statistical tables published by Statistics Iceland. *Population development in Iceland 1841–2066*. http://px.hagstofa.is/pxen/pxweb/en/Ibuar/Ibuar__mannfjoldi__1_yfirlit__ yfirlit_mannfjolda/MAN00109.px. Accessed September 14, 2018.

Statistics Iceland. (2018c). *Educational attainment*. https://statice.is/statistics/society/education/ educational-attainment/. Accessed April 29, 2018.

Statistics Iceland. (2018d). *Participation in lifelong learning continues to decrease*. https://statice. is/publications/news-archive/education/lifelong-learning-2017/. Accessed April 29, 2018.

United Nations. (2015). *World population ageing 2015*. New York: United Nations.

Warehouse of Opportunities. (2018). *Create a dynamic third age*. https://warehouseofopportunities. eu/. Accessed September 14, 2018.

Hans Kristján Guðmundsson is a Board member and former Chairman of U3A Reykjavík. After retiring he has engaged himself in research focusing on third age issues, both in international and European projects. Previously, he was Dean of the School of Business and Science at the University of Akureyri, Director General for RANNIS, Icelandic Centre for Research, and Rector for NorFA, Nordic Academy of Advanced Study, Oslo. A physicist with a TechnD. degree from KTH, Stockholm, he has extensive experience of international cooperation in research, research training, science policy and administration on Nordic and European level and especially in European Union programmes in research, education and culture.

Chapter 5
The University of the Third Age in Italy: A Dynamic, Flexible, and Accessible Learning Model

Barbara Baschiera

Introduction

According to the European Commission's Lisbon Strategy—issued in 2000, reiterated in Barcelona in 2002, and constantly reaffirmed thereafter in key policy documents—Italy and all the other European Union member states are committed to identify coherent strategies and practical measures to foster 'lifelong learning for all' (European Council, 2000). Nevertheless, regardless of the fact that the number of older adults in Italy, who are looking to acquire new knowledge and skills is growing at an unprecedented rate, the Italian government has been reluctant to recognise the need to provide educational opportunities for seniors. Indeed, it was only in the last few years that Italy formally recognised the benefits of education and learning programmes for adults. In this respect, different proposals have been made at national level: community education, adult education, and finally lifelong learning. Although proposals made so far have been conceived with specific objectives in mind, age-relevant educational targets still need to be established to-date. The concept of 'learning' itself has over time come to bear double meaning. On the one hand, it means learning at any age (*lifelong learning*) by extending to *lifewide learning* (where learning occurs in formal structures as well as through other experiences). On the other hand, it means *learning for all*—that is, trying to meet everyone's needs. As a result, it should not come as a surprise that attracting older adults back to education, requires policy makers to consider that lifelong learning is not made up exclusively of formal education circuits, but also includes opportunities from non-formal education (Istituto per lo Sviluppo della Formazione Professionale dei Lavoratori, 2003).

In Italy, the majority of non-formal learning activities for older adults are offered by the Universities of the Third Age (U3As) and, to a lesser extent, by volun-

B. Baschiera (✉)
University of Malta, Msida, Malta
e-mail: barbara.baschiera@um.edu.mt

© Springer Nature Switzerland AG 2019
M. Formosa (ed.), *The University of the Third Age and Active Ageing*, International Perspectives on Aging 23, https://doi.org/10.1007/978-3-030-21515-6_5

tary organisations. The term 'university' is used in the medieval sense, denoting member students engaged in the selfless pursuit of knowledge and truth as an end-in-itself. U3As constitute settings where older persons participate in cultural projects that augment both self-awareness and social responsibility. In Italy, there are five organisations which run recognised U3As: Unitre (250 branches) (http://www.unitre.net), FederUni (250 branches) (http://www.federuni.it), Uniauser (120 branches), Cnupi (40 branches) (http://www.cnupi.it), and Unieda (34 branches) (http://www.unieda.it) (Luppi, 2016; Principi & Lamura, 2004). Although all U3As share common objectives such as a lifelong learning ethos that considers some form of gerotranscendence as essential, there are also significant differences in structure and organisation between them; these vary according to the particular features of each U3A and the ways in which they are accessible at local level. This chapter charts the founding and development of U3As in Italy, focusing on theoretical learning perspectives, and members' profiles and motivation to enrol. It also addresses the preferred and most common curricula, and preferred teaching strategies. The chapter concludes by noting the contributions of U3As in Italy towards active and successful ageing, and the movement's most pressing challenges in the foreseeable future.

Origins and Development

U3As in Italy have developed differently from those in other European countries, where they have been promoted by regular universities (Principi & Lamuria, 2009). Such peculiar development, on the one hand, has facilitated local connection and focus on the users, whilst on the other hand has caused the development of precarious, transient, and different U3A models. In Italy, the term 'University of the Third Age' is actually an extensive term that describes different networks of educational providers, and a wide range of learning programmes and participants. The International Association of Universities of the Third Age (AIUTA) was founded in 1975 to connect the growing U3A movement (Formosa, 2014). Nevertheless, despite its international title it is only recently that AIUTA has recognised the validity of U3As which are not linked to traditional universities. The presence of AIUTA at international committees is a positive and strong strategy to put the issues of older adult learning on the political agenda by exerting pressure for more opportunities for older adult learning this also improving the well-being of older persons in general.

The first U3A in Italy was founded in Turin in 1975, followed by many others in the late 1970s and 1980s. During the latter period, U3A coordinators in Italy were aware that most U3As oscillated between two key models—namely, the francophone and anglophone models—and reacted by opting for a more culturally hybrid U3A model for the Italian context; this is one that links and combines an ethos for learning with sociocultural objectives, whilst embracing the principles of self-help and self-determination. Indeed, Italian U3As are not affiliated with, or administered by, regular or adjunct university faculties, nor are they established and funded by local government. Maintaining high standards is not the primary objective of U3As in

Italy, as curricula are not drawn up by university committees, and teaching is not facilitated by academics. Similar to the anglophone U3A model, Italian U3As include no 'top-down' administrative arrangement, but opt for a 'bottom-up' approach that exemplifies a culture of mutual aid and voluntarism. Following one of the Laslett's (1989) key principles, most of the activities at Italian U3As are run by volunteers and made freely available to all members. All U3As in Italy are independent, self-governing, and democratically run organisations and open to all without academic admission requirements or examinations. As recently stated by Luppi,

> Third age universities set out to promote the dissemination of culture, foster the inclusion of older persons in the social and cultural life of their local communities and provide appropriate responses to the educational and learning needs of citizens. The cultural activities and courses provided by third age universities generally cover a wide range of theoretical subjects and some practical activities (ranging from literature, theatre, visual arts, history, philosophy, psychology, religion, politics, economics, science, computing, foreign languages, music, choir, dance, expressive workshops including painting, embroidery, to restoration). The course catalogues are based on a careful analysis of the characteristics, needs and interests of the local community and the members of each university association.
>
> Luppi, 2016: 204

The most recent data on Italian U3As claimed that as much as '96.5% have external teachers, the majority of them (57.3%) being volunteers, although some are also hired as occasional consultants (31.3%, especially school teachers), and hardly ever as permanent staff' (Principi & Lamura, 2009: 252). The administration of U3As in Italy is generally based on volunteers who take on both management and teaching roles, whilst offices are run on minimal operational costs. The range of courses is extensive, and classes are conducted throughout the year. Only a few of them receive any assistance or financial support from local government, with U3A members paying between a 100 and 200 € per semester (12–14 weeks). This amount allows them to have access to the biweekly 2–3 h lectures and distinctive monthly seminars. Hence, the cost of lifelong learning for older Italian citizens is relatively affordable and allows for a significant range of participation. The U3A model fosters a system where learners also take on the role of teachers and each U3A member represents their own skills, expertise, and life experiences to others. Indeed, learning programmes tend to include intellectually, physically, and socially stimulating topics which are planned and coordinated by retired people who are knowledgeable in their subject areas.

Theoretical Learning Perspectives

In the light of the determination of older adults to participate in learning opportunities, it is crucial to find out which educational models best meet their interests and goals (Dal Ferro, 2009). Polverini and Lamura (2003) stated that geragogical learning models are still missing in Italy, making it difficult to effectively meet the diversity of needs expressed by senior citizens. However, since the early 2000s many theoretical perspectives have guided the implementation of learning programmes at U3As (Findsen & Formosa, 2011; Formosa, 2010a, 2016a). Whilst it is true that some facilitators still adopt a traditional (pedagogical) approach, which is based on an overly disciplined transmission of knowledge, an increasing number of facilitators are utilising a more participative approach. As Knowles, Holton, and Swanson (1998) highlighted with respect to adult learners, older adult learners are also highly independent in their learning approach and prefer to take an active role in the class without following any rigid teaching schedule. Older learners value a tutor who is interested in their opinions, who treats them as peers and gives them opportunities to express their opinions. As a result, many facilitators follow a liberal learning paradigm to consult older adults on matters related to the curricula, teaching pattern, materials for learning, and seating style. Older learners like to set their own goals, and these might differ even in the same class. Many tutors at Italian U3As disclose to learners their aim and objectives early on in the training programmes, whilst also encouraging and assisting learners in setting personal goals by establishing personal targets at the very first session.

Since learners are generally weighing reasons to enrol in a specific learning programme and exploring and evaluating specific programmes for their relevance to their interests and plans, many tutors address these matters at the beginning of sessions, especially when new topics are introduced, and they also ensure that concepts are presented in contextual settings that are familiar to older persons. Another approach that has guided the implementation of learning strategies at U3As is founded on the belief that the learners' lifelong and lifewide experience should have a central role in the learning process. As Knowles et al.'s (1998) work emphasised, mature learners bring rich life experience to the class, an experience that emanates from their occupation careers or even from family experiences or life transitions. U3A tutors are generally ready to recognise when learners are drawing on their life experience, and quickly act to connect the ongoing learning process and outcomes derived from it. When this occurs, the role of older learners at U3As is not to sit perfectly still, listening passively, but actually to have an active role in the learning experience. In order to stimulate such an 'active' form of participation, facilitators tend to utilise role-playing or discussion techniques which encourage learners to forward their views on the subject. When older adult learners can connect new learning to their own experience through a meta-cognitive process, the taught material is more likely to be understood and remembered. By reflecting on the ongoing learning process in terms of their own experience, older adults are able to transform their earlier experience into a deeper understanding of both the surrounding natural and social environment (Kolb, 1984).

Members Profile, Motivation, and Curricula

Although U3As in Italy are open to people of all ages, the average age of members is between 65 and 75, the majority of whom are female (the ratio is about 2 females to 1 male). This may be due to different reasons such as the gender differences in life expectancy in Italy, and the social and informal character of the U3A learning experience which tends to be more attractive to older women compared to their male counterparts. As Williamson (2000) noticed, older men and women approach learning in later life in different ways: men in U3As prefer to 'sit and think', whilst women like being more active and 'free' to do things that they were unable to perform previously. Moreover, the 'open door' policy means that U3As attract participants from various socio-economic backgrounds, attracting retirees with higher-than-average educational qualifications who held leadership positions in their former occupation roles, but also peers with no or basic schooling experience who wish to update their knowledge and skills in a variety of subjects. The search for new opportunities to socialise the desire to develop one's own psychophysical independence and the ambition to achieve a higher cultural and scientific knowledge is amongst the main motivations for older adults in Italy to enrol in U3As. In other words, the social aspect of learning seems to be the key reason for participating in learning avenues, spurred by emergent social bonds from meeting same-aged peers, escaping daily routine, making new friends, and joining a community of learners. Older learners' interests vary greatly regarding contents and subject areas. Programmes reflect the needs and interests of older learners through age-relevant curricula. Courses vary widely in contents and formats: study or discussion groups, workshops, excursions, seminars on topics such as physical exercise, health and self-care, human relations, social conflict, creative and biography writing, courses about the computer and the Internet, social sciences, women's issues, and local and general history. Art is often taught through artistic experiences such as drawing, painting, theatre, and music. Principi and Lamura's study found that in Italian U3As the

> …prevocational courses are, in general, not as popular, the most attended ones in this category being those on ceramics, restoration, painting, sculpture, mosaics, glass work, and arts in general. The highest percentage of participants is in…general culture, sport activities, expressive education (graphics, plastics, literature), cinema language, photography, and music are the most popular. As far as the percentage of over 65 participants at the various courses is concerned, the courses with a high percentage of senior citizens are those on gardening (50%), voluntary work education (48.5%), computer science and Web design (47.4%), education on the rights of the citizen (43.2%), musical education (42.7%), general culture (41.2%), and environmental education (41%).
>
> Principi & Lamura, 2009: 250–1

As expected, classes are heterogeneous in terms of participants, but facilitators attempt to provide U3A members with the best possible learning experience, whether it involves learning new skills, a social or cultural experience, or simply as a source of inspiration and enjoyment.

Learning and Teaching Strategies

One key challenge for U3As involves the setting of learning and teaching strategies that are attuned to all members irrespective of their socio-economic backgrounds and lifetime experience (Formosa, 2012). Moreover, older adult learners expect facilitators to treat them as peers, to allow them an exchange of opinions, and not to digress from the established learning outcomes. Jarvis (2001) observed that new approaches, as opposed to traditional education models, were required for late-life learning to be a success. In Italy, the learning programmes, formal lectures, discussion groups, or seminars organised by U3As are attended by a number of participants varying from groups of around 15 people to ones of around 100, depending on the subject and the geographical setting. The teaching and learning strategies at U3As are aimed at creating a welcoming approach and a positive atmosphere so that all learners feel comfortable and socially included. Facilitators coordinate training programmes in a way that learners are challenged but not to the extent that they feel discouraged, stressed, or bored. They encourage learners to ask questions, ask for clarifications, and to regard challenges as positive experiences. This hermeneutical approach, which aims to understand rather than illustrate, helps to contextualise learning and applies it to real life. Other facilitators who may opt for a more experience-based learning approach include structured activities such as simulations, games, role plays, group discussions, and/or acting. In such circumstances, facilitators coordinate case studies, role plays, video-based activities, group discussions, autobiographical writing, problem-based learning, group work, and self-directed projects. Briefly, experience-based learning draws on the learners' life experience, as it engages the whole person, stimulates reflection on experience, and encourages openness towards new experiences and, overall, encourages lifelong learning. As expected, peer learning is popular amongst facilitators and members alike, as in Findsen and Formosa's words

> ...older persons find peer teaching as a way to meet their need to feel needed. Peer teaching is a learner-centred activity because members of educational communities plan and facilitate learning opportunities for one other. Peers will plan and facilitate courses of study and be able to learn from the planning and facilitation of other members of the community: "peer teaching is a rare and provocative model of education in which, in the morning, a person may teach a class for her peers, and that same afternoon have one of her 'students' become her teacher"...

> Findsen & Formosa, 2011: 104–5

Facilitators also tend to go beyond descriptive explanations, and instead, prefer shared research, analyses, and interpretation of cultural facts where learners are given the role of active participants. Assessment is excluded in U3A classes because of the 'voluntary learning' principle, and the fact that people attend training programmes for the sake of learning as an end-in-itself. Instead of promoting learning in isolation, the U3A model encourages participants to share their knowledge collectively, integrating everybody's ideas and talents into a democratic learning design. Therefore, the co-construction of learning experiences through dialogue, research, and practice benefits from its strong grounding in autonomous and democratic learning

(Findsen, 2006). Facilitating sharing at U3As aims to develop confidence, as older adults appreciate the value of the knowledge they share with others. The facilitation process also supports the attitudes, previous learning and world views that learners bring to the class, acknowledging and giving value to every member's experience and contribution. Facilitators thus embody an ethical stance that accords respect to the learners, values each individual, and recognises their need for self-direction. Follow-up activities, self- and peer-assessments are carried out to provide regular critical feedback. As facilitators are provided with freedom in the way they coordinate their training programmes, it results that the latter may differ from a very structured setting to a very informal scenario where facilitators actually consider themselves as other members of the class. However, they remain at all times responsible for the direction of the class:

> Educators therefore hold a position of authority deriving from his/her competence which, in turn, commands, respect. Authority must, however, never degenerate to a form of authoritarianism since 'the educator's task is to encourage human agency, not moulded in the manner of Pygmalion'…
>
> Formosa, 2011: 327

In brief, older adult learning is very different from learning at universities and other tertiary institutions. The facilitators' approach is generally holistic, implying that they must recognise and cater for the human needs of learners, which can include educational, cultural, and social elements.

Implications for Active Ageing, Quality of Life, and Well-Being

U3As in Italy are playing an important, albeit largely unrecognised role in the country's active ageing process, as they are addressing a specific audience of adults who are motivated to improve their quality of life and well-being. The emergent experiences provide Italian older persons with an opportunity to engage in social participation, cultural promotion, and active citizenship, in the face of widespread ageism and age discrimination. In this way, U3As allow older adults to fulfil their potential—mostly as the result of statutory mandatory retirement—and provide them with the opportunity to stay physically, socially, and mentally active, as well as strengthen national levels of collective memory and intergenerational solidarity (McQueen, Hallam, Creech, & Varvarigou, 2013). U3As allow older adults to experience the joy of acquiring and formulating new knowledge, having a positive effect on older persons' lives. One key impact is that people who were previously not able to engage in formal educational programmes because of either work or family commitments now have easy access to cultural learning initiatives. Moreover, U3As can facilitate communication skills and critical thinking amongst older persons, which in turn fosters levels of social inclusion in later life, as well as stimulating creativity and new ways of expression. Many studies have, in fact, shown the various benefits of adult learning programmes that include opportunities for physical and cognitive energy, recreation and social interaction for the overall health of the elderly, bringing advantages such as increased intellectual development, life satisfaction, personal fulfilment, creative expression, and social networking building (Bunyan & Jordan, 2005; Escolar Chua et al., 2014; Formosa, 2013; Hafford-Letchfield, & Formosa, 2015; Hebestreit, 2008; Siedle, 2011). Indeed, Formosa's perceived benefits of U3As for members resonate extremely well with the Italian context:

> [U3As] fulfil various positive social and individual functions such as aiding lonely older persons to re-socialize themselves by enabling them to form new groups and increase their interests. They also provide opportunities, stimulation, patterns, and content for the use and structure of the older persons' free time which would otherwise be characterised by inactivity. U3As also develop in members a lofty and progressive delight of life, increase the social integration and harmony of older persons in society, inject a sense of creativity in older persons, and make older persons more visible in society...U3As also address various intellectual, emotional, physical, leisure, and spiritual needs of older persons, as well as provide older persons with the opportunity to organise and coordinate social/cultural activities and thus make their lives more fruitful and energetic.
>
> Formosa, 2010b: 5

Indeed, successful participation in educational and learning programs has the potential to support a feeling of self-efficacy, especially since being part of a U3A group may improve members' levels of self-confidence. Thanks to the development of the ability to access information, U3As can assist older people to become more aware of their human and legal rights and thus mitigate against elder abuse and foster active citizenship in both public and political spheres. As governments everywhere are looking for inexpensive solutions to challenges associated with population ageing, investing in U3As is surely a relatively efficient and uncomplicated way to augment the levels of active, healthy, and positive ageing.

Conclusion: Looking Towards the Future

The continuous ageing of Italian society throughout the current century (Italy is the second oldest country in the world) suggests that more older adults will join U3As in the near future and thus will require more varied continuing educational programmes, putting on more logistic and financial pressure on the U3As. Italian administrators are already reporting difficulties in recruiting course directors and facilitators/tutors. Should the number and type of lectures be required to increase, funding may become even more of a challenge. There is a vast difference between informal and formal provision in terms of accessible state funds which are, truth be told, non-existent for U3As. Unfortunately, even European Union funds that are allocated to lifelong learning projects are used to redevelop workforce, productivity, and the economy rather than to enhance learning for ageing and retired adults (Borg & Formosa, 2016; Formosa, 2016b). The fact that U3As rely on their members' voluntary work may actually be turning against the interests of this movement, as governments tend to take this global and national initiative for granted. Another pressing challenge is that not all volunteering facilitators have the required geragogical skills to coordinate learning programmes for older adults. Since the teaching and learning of older adults requires a specific *raison d'être*, facilitators need not only to be prepared academically, but also to be trained socially and culturally about a number of issues concerning older learners, to be able to creatively modify the conventional provision of education and to offer them those opportunities that cater to their physical, psychological and social needs. Perhaps quality, reliability, and consistency of the learning provision for older adults might be best established through professional training courses. The relative absence of men amongst the U3As' members remains a challenge as fewer men are attracted to seek membership (Formosa, Chetcuti Galea, & Farrugia Bonello, 2014; Formosa, Fragoso, Jelenc Kraŝovec, & Tambaum, 2014; Formosa, Fragoso, & Jelenc Kraŝovec, 2014). To this effect, it would be best if U3As take a leaf out of the Men's Sheds' movement and realise that

> …the most effective learning from men in community settings occurs where learning intentions are not formalized or brought to the fore, where the pedagogies build on what men know, and where social relationships rather than courses or enrolments are emphasized. For men with the most negative attitudes towards learning, pedagogies based on communities of men's informal practice have been found to be effective.
>
> Golding, 2012: 144

Moreover, if the U3A wants to become a more inclusive learning model it needs to attract other disadvantaged population groups—such as older persons with disability, ethnic minorities, and residents in long-term care facilities—whilst also revising the concept of learning as a formal classroom-based activity in favour of other alternative styles and settings. It is also augured that in the near future Italian U3As will implement online learning procedures, creating online platforms for older people to share their life experience, skills, and talents:

> U3As must not assume that older learners continue living in some by-gone world. Rather, e-learning has become increasing popular in later life as it offers the opportunity for older learners to access information and communicate with others when and if they want to. For U3As to continue being relevant to contemporary elders, centres must make more effort to embed their learning strategies in the web 2.0 revolution that now provides extremely user-friendly applications…[E-learning] offer[s] limitless possibilities for an interactive, empowering, and participatory form of older adult learning.
>
> Formosa, 2014: 48

E-learning represents an accessible, comfortable, safe environment to encourage learning at an appropriate pace, suitable to the needs of the participants involved. It enables older learners to be in charge of their own learning experience, deciding what to learn, how to learn, with whom to learn, what goals to achieve, and what value they get from learning (Talmage, Lacher, Pstross, Knopf, Burkhart, 2015). Nevertheless, despite the challenges that U3As in Italy are currently facing—and will be facing more starkly in the foreseeable future—the movement remains a dynamic, flexible, and accessible learning model for older Italian adults. The movement changes contours whilst remaining within the spirit of the original principles, and this is testament to its dynamicity; its chameleon approach to teaching, learning and curricular models is witness to its flexibility in the provision of subject modules; whilst the fact that its premises are located within various structures in both rural and urban locations is proof of its accessibility to as wide a range of older adults as possible.

References

Borg, C., & Formosa, M. (2016). When university meets community in later life: Subverting hegemonic discourse and practices in higher education. In J. Field, B. Schmidt-Hertha, & A. Waxenegger (Eds.), *Universities and engagement. International perspectives on higher education and lifelong learning* (pp. 105–116). London: Routledge.

Bunyan, K., & Jordan, A. (2005). Too late for the learning: Lessons from older learners. *Research in Post-Compulsory Education, 10*(2), 267–281.

Dal Ferro, G. (2009). *Insegnare agli adulti: Note di metodologia e didattica per le Università della terza età*. Vicenza: Edizioni Rezzara.

Escolar Chua, R. L., & de Guzman, A. B. (2014). Effects of third age learning programs on the life satisfaction, self-esteem, and depression level among a select group of community dwelling Filipino elderly. *Educational Gerontology, 40*(2), 77–90.

European Council. (2000). *Lisbon European Council—Presidency Conclusions*. http://www.consilium.europa.eu/uedocs/cms_data/docs/pressdata/en/ec/00100-r1.en0.htm. Accessed June 12, 2018.

Findsen, B. (2006). Social institutions as sites of learning for older adults: Differential opportunities. *Journal of Transformative Education, 4*(1), 65–81.

Findsen, B., & Formosa, M. (2011). *Lifelong learning in later life: A handbook on older adult learning*. Rotterdam: Sense.

Formosa, M. (2010a). Universities of the Third Age: A rationale for transformative education in later life. *Journal of Transformative Education, 8*(3), 197–219.

Formosa, M. (2010b). Lifelong learning in later life: The Universities of the Third Age. *Lifelong Learning Institute Review, 12*, 1–12.

Formosa, M. (2011). Critical educational gerontology: A third statement of principles. *International Journal of Education and Ageing, 2*(1), 317–332.

Formosa, M. (2012). Education and older adults at the University of the Third Age. *Educational Gerontology, 38*(1), 114–125.

Formosa, M. (2013). Creativity in later life: Possibilities for personal empowerment. In A. Hansen, S. Kling & Ŝ. Strami González (Eds.), *Creativity, lifelong learning and the ageing population* (pp. 78–91). Ostersund, Sweden: Jamtli Förlag.

Formosa, M. (2014). Four decades of Universities of the Third Age: Past, present, and future. *Ageing & Society, 34*(1), 42–66.

Formosa, M. (2016a). Malta. In B. Findsen & M. Formosa (Eds.), *International perspectives on older adult education: Research, policies, practices* (pp. 161–272). Cham, Switzerland: Springer.

Formosa, M. (2016b). Critical friend commentary. (not) Talking about my generation: Is higher education engaging older learners. In G. Steventon, D. Cureton, & L. Clouder (Eds.), *Student attainment in higher education: Issues, controversies and debates* (pp. 188–189). New York: Routledge.

Formosa, M., Chetcuti Galea, R., & Farrugia Bonello, M. (2014). Older men learning through religious and political affiliations: Case studies from Malta. *Androgogic Perspectives, 20*(3), 57–69.

Formosa, M., Fragoso, A., & Jelenc Kraŝovec, S. (2014). Older men as learners in the community: Theoretical issues. In S. Jelenc Kraŝovec & M. Radovan (Eds.), *Older men learning in the community: European snapshots* (pp. 15–28). Ljubljana: Ljubljana University Press.

Formosa, M., Fragoso, A., Jelenc Krašovec, S., & Tambaum, T. (2014). Introduction. In S. Jelenc Krašovec & M. Radovan (eds.), *Older men learning in the community: European snapshots* (pp. 9–13). Ljubljana: Ljubljana University Press.

Golding, B. (2012). Men's learning through community organisations: Evidence from an Australian Study. In M. Bowl, R. Tobias, J. Leahy, G. Ferguson, & J., Gage (Eds.), *Gender, masculinities, and lifelong learning* (pp. 134–146). Abingdon: Routledge.

Hafford-Letchfield, T., & Formosa, M. (2015). Mind the gap! An exploration of the role of lifelong learning in promoting co-production and citizenship. *European Journal for Research on the Education and Learning of Adults, 7*(2), 237–252.

Hebestreit, L. (2008). The role of the University of the Third Age in meeting needs of adult learners in Victoria, Australia. *Australian Journal of Adult Learning, 48*(3), 547–565.

Istituto per lo Sviluppo della Formazione Professionale dei Lavoratori. (2003). *Formazione permanente: Chi partecipa e chi ne è escluso. Primo rapporto nazionale sulla domanda.* Catanzaro: I libri del Fse.

Jarvis, P. (2001). *Learning in later life: An introduction for educators and carers.* London: Kogan Page.

Knowles, M. S., Holton, E. F., & Swanson, R. A. (1998). *The adult learner.* Houston, TX: Gulf Publishing.

Kolb, D. (1984). *Experiential learning: Experience as the source of learning and development.* Englewood Cliffs, NJ: Prentice Hall.

Laslett, P. (1989). *A fresh map of life: The emergence of the third age.* London: Weidenfeld & Nicholson.

Luppi, E. (2016). Italy. In B. Findsen & M. Formosa (Eds.), *International perspective on older adult education: Research, policies, practice* (pp. 201–210). New York: Springer.

McQueen, H., Hallam, S., Creech, A., & Varvarigou, A. (2013). A philosophical perspective on leading music activities for the over 50s. *International Journal of Lifelong Education, 32*(3), 353–377.

Polverini, F., & Lamura, G. (2003). La qualità della vita in età anziana: L'evidenza empirica in Italia. In L. Frey (Ed.), *Le condizioni di vita degli anziani in Italia* (pp. 7–83). Milan: Franco Angeli.

Principi, A., & Lamura, G. (2004). *National report on Italy.* Maastricht, The Netherlands: Pan European Forum for Education for the Elderly (PEFETE).

Principi, A., & Lamura, G. (2009). Education for older people in Italy. *Educational Gerontology, 35*(3), 246–259.

Siedle, R. (2011). Principles and practices of mature-age education at U3As. *Australian Journal of Adult Learning, 51*(3), 566–582.

Talmage, C. A., Lacher, R. G., Pstross, M., Knopf, R. C., & Burkhart, K. A. (2015). Captivating lifelong learners in the third age: Lessons learned from a university-based institute. *Adult Education Quarterly, 65*(3), 232–249.

Williamson, A. (2000). Gender issues in older adults' participation in learning: Viewpoints and experiences of learners in the University of the Third Age (U3A). *Educational Gerontology, 26*(6), 49–66.

Dr. Barbara Baschiera received her Ph.D. from Ca' Foscari University of Venice with the title of her dissertation being 'Beyond the age: Developing the educational potential of the older adults'. She is an academic at the University of Malta and carries out research on lifelong and intergenerational learning. Recent publications include *Intergenerational entrepreneurship education: Older entrepreneurs reducing youngsters' social and work disengagement* (2018) and *Key competences for lifelong learning in the education of seniors* (2018).

Chapter 6
Subsisting Within Public Universities: Universities of the Third Age in Germany

Bernhard Schmidt-Hertha

Introduction

With changing images of ageing and the discourse on active ageing, participation in adult education seems to become ever more popular among German older persons. Participation rates in continuing vocational education and training have increased significantly among older workers and are comparable to the figures for younger citizens, except for the last few years before retirement (Schmidt, 2007). As regards the field of non-vocational adult education, one out of ten older adults—but up to the age of 80—participates at least once in adult education within a period of 12 months; the same holds true for younger adults (Tippelt, Schmidt, Schnurr, Sinner, & Theisen, 2009).

In Germany, most of the older learners attend courses offered by adult education providers such as the *Volkshochschule* (Adult Education Centre), and only a minority of them enrol at Universities of the Third Age (U3As). This may be due to the fact that U3As in Germany are usually integrated into the public universities and are thus part of the higher education system. Accordingly, they offer almost exclusively academic courses at the level of bachelor or master studies, even for older adults. In addition, some university programs for older citizens require a university entrance qualification, which only 10% of older (65-plus) German citizens older hold (Statistisches Bundesamt, 2017). Overall, barriers to participation in U3A courses may be higher in Germany than in other countries due to formal requirements and the image of the U3A as an academic institution. In contrast to many other countries (Formosa, 2012, 2014, 2016), U3As in Germany are, in most cases, not independent institutions, but part of public universities and predominantly financed by the universities' budgets and participation fees. Thus, in the German federal system U3As are not controlled by the individual states' Ministries of Education, but by their Ministries

B. Schmidt-Hertha (✉)
University of Tuebingen, Tuebingen, Germany
e-mail: bernhard.schmidt-hertha@uni-tuebingen.de

© Springer Nature Switzerland AG 2019
M. Formosa (ed.), *The University of the Third Age and Active Ageing*, International Perspectives on Aging 23, https://doi.org/10.1007/978-3-030-21515-6_6

of Science. These structures lead, on the one hand, to a high academic level of the courses offered by German U3As and, on the other hand, to a marginalisation of the U3As in many universities, since they usually do not train young researchers and are not financed by the state as opposed to the degree programmes. However, due to these structures U3As can actually profit from the infrastructure provided by the universities and from their close proximity to research. In the chapter's subsequent sections, the history and the structures of U3As in Germany will be presented in depth and a close look will be taken at the composition of participants with their diverse needs and at the most popular didactical principles, which can be seen both as a reaction to older persons' needs as well as the institutional structures. Finally, the possibilities of intergenerational learning at German universities are also discussed.

History of U3As in Germany

As goes for many other European countries, the German U3A has its roots in a conference held at Nancy in 1979, where the idea of a university for people of the third age was presented first (Swindell & Thompson, 1995). With this idea in mind, the first international U3A workshop in Germany was organised at Oldenburg that same year. Since many of the universities participating voiced a strong interest in the concept, more workshops on the topic were organised by different German universities during the subsequent years. At that time, three central aims drove the process of establishing U3As. First, the educational policy of using higher education as a way of reducing social inequality and of creating new possibilities for social mobility, which prevailed during the 1960s and 1970s, lead to the idea of opening up universities to educationally disadvantaged groups. Second, the era was characterised by huge societal challenges and universities became a place to deal with those topics. And third, U3As were seen as an instrument for regional development, especially in economically underdeveloped regions. In accordance with these ideas, U3A programmes in Germany still have a scientific character, a close relation to research, and are open to all groups of people—independent of age or level of education—at least in most states (Bertram, Dabo-Cruz, Pauls, & Vesper, 2017; Dabo-Cruz & Pauls, 2018). During the 1980s and 1990s, U3As spread very quickly throughout Germany, and by 1996 as many as 35 U3As were active in Germany (Saup, 2001). In 1985, a *Federal Workgroup for Universities for Older Adults* was founded with representatives from 27 institutions. In 1994, the workgroup changed its name to *Federal Workgroup on University-Based Continuing Education for the Elderly* (BAG WiWA) and became part of the *Workgroup for Adult Education in Universities* (AUE), which today is called the *German Association for University-Based Continuing and Distance Education* (DGWF). BAG WiWA is still active and holds conferences and meetings every year (Bertram et al., 2017). In 2006, the first and only independent U3A—the *European Centre for University Studies for Seniors East-Westphalia*—was founded. It is the only U3A in Germany not linked to a public university.

Institutional Structures

In general, U3As in Germany are part of the public universities, and most of them were founded during the 1980s and 1990s. The predominant designation for these institutions is *Seniorenstudium* (Study Programme for Senior Citizens). However, other labels such as *Kontaktstudium* (Contact Study Programme), *Studieren 50plus* (Studying 50plus), or *Studium für Ältere* (Study Programme for Older Persons) can be found (Schmidt-Hertha, 2018). Some of these programmes work under the auspices of the universities' continuing education departments, while others are organised in separate centres within the universities, carrying names that do not explicitly refer to older adults (e.g. University of Lifelong Learning or Centre for General University-Based Continuing Education). Several of these programmes aim at preparing older learners for civic engagement and honorary offices. In this sector, three different models of German U3A programmes can be distinguished (Dabo-Cruz & Pauls, 2018).

- First, there are programmes training older adults for civic engagement or for becoming teachers for other older adults (e.g. Begemann, 2009). These programmes are driven by the requirements of jobs that are or will be carried out by the older learners and that may well be the main reason for them to participate in the programme.
- Second, we have well-structured programmes with a focus on general education, often referring to central challenges of society (e.g. social inequality, sustainability, etc.). In contrast to the first model, the curricula are not chosen to prepare learners for specific tasks but to follow the general idea of *Bildung* (Education), aiming at a better understanding of oneself and the world around us (Marotzki, 1990).
- Third, there are selected courses within degree programmes which are open to U3A students. They seem to be particularly interested in subjects offering another perspective on the individual, culture, and society, e.g. theology, philosophy, psychology, history, history of arts, and social science (e.g. Kaiser, 2009; Kröll, 2011). This model seems to be quite resource-friendly, which may explain why it is the most common model among German U3As (Bertram et al., 2017). It also brings together older and younger students, with all the potentials and risks inherent in this kind of intergenerational learning at universities (Stahl, 2011; Tremmel, Bschaden, & Wagermaier, 2014)—a topic that will be discussed later.

At the time of writing, many German universities provide such programmes for older students, and some even combine two of these three models. The majority of the U3As are members in the BAG WiWA, which currently has 59 institutional members, including different types of universities located in Germany, as well as two universities from Austria and one in Switzerland. The network is engaged in strengthening the lobby for older learners at the university level and has formulated five central demands on higher education policy in Oldenburger Erklärung (2018) (Oldenburg Declaration). In this statement, members of BAG WiWA demanded (i) respect for the positive effects of learning in later life, (ii) the expansion of educational offers for older adults and of intergenerational programs, (iii) support for university-based continuing education institutions promoting learning in later life, (iv) additional financial resources for university-based continuing education, and (v), engagement in promoting inclusive education which also offers learning opportunities for older adults. These organisational structures suggest that there is a relatively large community of older learners in the universities. However, this is not the case. According to a survey carried out by Tippelt et al. (2009), only one out of 100 educational activities in U3As is actively planned and carried out by Germans aged in the 65–80 age bracket. Hence, in comparison with many other countries U3As in Germany only play a marginal role in the market for older adults' education programmes. One reason for this could be the self-conception of many German universities, which do not consider continuing education one of their central tasks, but rather see it as a by-product which does not contribute significantly to their reputation. Further reasons for the overall marginal share of universities on the German adult education market can be found in the structures of university-based continuing education mentioned above (Feld & Franz, 2016). Even though continuing education is defined as one of the tasks of universities in the higher education acts of almost all of the German states, there is rarely any sufficient sustainable funding for universities to fulfil this task.

Participants

University-based continuing education offers opportunities to expand professional expertise (Sagebiel, 2014) or to catch up on missed educational qualifications (see also Field, Schmidt-Hertha, & Waxenegger, 2015). The latter is particularly relevant for many older women, who often did not have the chance to acquire higher education qualification in their younger years. Being active as a U3A teacher also offers older adults the possibility to make use of and share their professional knowledge and skills even after the end of gainful employment. At least 11% of all adult educators in Germany are older than 65 (Martin, 2016: 65). At first glance, adult education statistics in Germany seem to show that participation in learning activities is clearly associated with age (Bundesministerium für Bildung und Forschung, 2015). However, this correlation decreases once labour status is taken into account, as being retired also means losing access to continuing vocational training programmes provided by entrepreneurial enterprises. Once further variables are controlled for, some studies find only marginal effects of age on participation (Schmidt, 2007; Aust & Schröder, 2006). Factors other than age seem to be of much greater relevance in predicting educational behaviour, for example, labour status, level of formal education, or images of ageing (Schmidt-Hertha & Mühlbauer, 2012). For all age groups, an extended participation rate in adult education can be shown for people with a university degree (Tippelt et al., 2009). This means that for German adults aged 65-plus only one out of four older adults holding a university degree is engaged in adult education within a twelve-month period, an unflattering statistic when one considers that only one out of ten older persons in Germany holds such a formal qualification (ibid.).

Using public data on demographics in Germany and results from Tippelt et al. study (2009), it can be calculated that about 666,000 adults aged 65-plus are engaged in some kind of adult education programme within the course of a given year, and about 60,000 of them participate in U3A courses. However, it has to be noted that the majority of the U3A students have not reached the statutory retirement age of 65 yet, and more than one-third of them are between 61 and 65 years old, while another 20% is aged 60 years or younger (Sagebiel & Dahmen, 2009). Indeed, even though most states do not demand a university entrance qualification for U3A learners, most of the U3A students do have one (Tippelt et al., 2009). Considering that the future generations of older adults in Germany will be even larger and will have a higher share of people with some form of graduate qualification, a further increase in the number of U3A students in Germany is to be expected as long as universities' doors remain opened to senior citizens. According to a survey of 1,110 German U3A students (Sagebiel & Dahmen, 2009), maintaining cognitive skills and abilities is the most common reason for participating in U3A programmes—mentioned by 84%—followed by the opportunity to finally study what the respondent had always been motivated to study (58%) but never had time for due to parenting and occupational responsibilities. Social aspects—such as getting in touch with other older learners and learning together with younger students—are mentioned only by 32 and 21%, respectively. Another reason for participating in U3A programmes mentioned by almost one out of four respondents was the opportunity to engage in scientific debates, which again shows that U3As are often associated as being a scientific/educational space.

Didactics

Since older adults differ in their learning requirements, educational aims, and reasons for learning, their expectations and demands regarding U3A programmes are highly diverse. Taking into account that older adults with higher qualifications, in particular, are used to learning continuously throughout their life, it is not surprising that didactic concepts with a focus on learner-control and with high demands on learning abilities seem to be very successful at German U3As (Costard, Haller, Meyer-Wolters, & Pietsch-Lindt, 2012; Malwitz-Schütte, 2000; Stadelhofer, 2000a, 2000b). Two didactical strategies—research-based learning (Costard et al., 2012; Stadelhofer, 2000a) and self-regulated learning (Kraft, 1999)—have proven to be highly productive in U3A programmes. Both strategies build upon explorative and autonomous learning and can be seen as appropriate concepts for experienced learners in learning arrangements that are not so much driven by preparing for an exam, but by the learners' interest in the topic. There is a huge body of international research on self-regulated (Pintrich & de Groot, 1990), self-organised (Low & Jin, 2012), or self-directed learning (Brookfield, 2009) in late-life learning, showing that these kinds of learning scenarios can be successful in very different contexts (school, vocational training, university, etc.) and for a broad range of curricular contents. However, this does not hold true for the idea of research-based learning, which seems to be appropriate for universities, in particular, which in general consider research one of their central tasks (Tremp, 2015). This is the case because research-based learning requires not only resources such as scientific literature or laboratories, but also training in research methods for the learners—resources to generally found first and foremost at universities. Given their solid research skills and access to necessary resources for research, such learners can develop their own research questions, start their own projects, and finally write a report on their research work. This scenario goes together with collaborative learning under professional supervision (Green, 2016), but can be detached from lectures offered at the U3A. As has already been shown by Stadelhofer (1999, 2000a, 2000b) and Malwitz-Schütte (2000), U3A programmes can unfold high and constructive momentum if older learners are given space to organise their learning on their own and to generate new knowledge.

In order to be successful, these scenarios require not only a high level of learning competence on the part of the participants, but also professional learning-support and guidance from the university staff (Kraft, 1999). In other words, not all U3As may be able to realise these demanding didactical concepts, as they require the appropriate organisational environment and resources. Since U3As in Germany are usually integrated into public universities, which traditionally combine research and teaching, they are in a way predestined to offer this kind of organisational environment. Nevertheless, these innovative didactical scenarios may not offer the adequate learning method for all older adults. Many participants in U3A programmes prefer to just attend lectures or discuss topics they are interested in with other learners and experts during seminars. Even though there are only very few studies available, it can be assumed that lectures and seminars are still the most common way of learning at German U3As. According to the survey by Sagebiel and Dahmen (2009), 8% of the U3A participants were involved in some kind of research-based learning and 2% in action research programmes. In addition, 10% reported to have participated in research travelling and 16% were involved in larger research projects. Self-organised learning, however, seems to be more popular among German U3A students than research-based learning. A total of 43% participated in courses or events organised by U3A students themselves (ibid.).

Another didactical concept applicable to U3As which seems to be not so popular is the idea of intergenerational learning (Schmidt-Hertha, 2014a; Schmidt-Hertha, Jelenc Krašovec, & Formosa, 2014). Being part of public universities, German U3As would have a great many opportunities to foster intergenerational learning (Ladas & Levermann, 2007; Schmidt-Hertha, 2017), but so far this concept has rarely been used as a strategic element. Making degree programmes available to U3A students—a practice widespread among universities—has the potential to bring together older and younger students (Steinhoff, 1997). Yet, these scenarios are often discussed as a source of conflict between younger students who aim at 'passing' an exam and older students who follow their personal interests and do not wish to be subject to the pressure of having to prepare themselves for the labour market (Gösken, 2012; Seidel & Siebert, 1998). Considering their life experience and vocational experience, U3A members possess a lot of knowledge that could be of interest to younger students and, in turn, these younger students have different skills and ideas which could be inspiring for all U3A participants, especially with regard to the use of digital media. Currently, digital media are seen as an important driving force for the future development of universities (Hochschulforum Digitalisierung [Higher Education Forum Digitalization], 2016), and there are many political activities and funding programmes supporting schools, universities, and adult education providers in developing new digital learning environments. With regard to U3As, too, digitalisation could open up new possibilities and create new potential, for example, by acquiring new target groups through innovative distance education programs, MOOCs (Massive Open Online Courses), or similar formats. Older adults in rural areas, with limited mobility, or with health problems could thus attend U3A courses without having to be regularly present. It goes without saying, nevertheless, that these digital learning environments demand a certain level of media literacy on the part of the learners, a level which may be held by many older adults compared to younger peers (Schmidt-Hertha, 2014b). It is noteworthy, on the other side of the coin, that the percentage of Internet users among older adults, and in particular among those with higher levels of education, is increasing steadily: 75% of Germans aged 60-plus use the Internet at least occasionally, and 44% use it on a daily basis (Koch & Frees, 2017).

The Future of U3As in Germany

U3As in Germany are widespread in the public university sector, but they are often organised as small units caring for only a small number of students compared to the huge number of young students enrolled in the degree programmes. So far, these structures have worked well because many older adults did not participate in adult education at all and those who did only rarely went to the universities. However, taking into account the continuously increasing number of people in their post-retirement phase of life, increasing participation rates of older adults in adult education, and a higher level of formal education among the coming generations of senior citizens, a growing number of U3A students are to be anticipated. More professionalised organisational structures with their own budget and staff could be one future model able to tackle these challenges, with such structures already in place in a small number of universities. With the growing number of older students, U3As can no longer be just an additional job for some professors or administrative staff who look for courses in degree programs to be opened up to U3A students, a practice which has anyhow become more difficult since the Bologna Process (Sagebiel, 2006), but which is still widespread among German universities. For U3As to become an important actor in the field of older adults' education, universities have to be given the necessary resources and landmark decisions have to be taken by political decision-makers clarifying what the task of universities should be in the field of learning in later life. At the same time, grass-roots movements like initiatives by the senior students themselves could be fruitful avenues to speed up the political process (Dummer, 2007). Thus, the *European Federation of Older Students* (2018), founded in 1990, makes it possible for senior students from all over Europe to organise themselves to voice their interests and their right to accessing higher education, a voice which certainly will become louder as their number multiply.

References

Aust, F., & Schröder, H. (2006). Weiterbildungsbeteiligung älterer Erwerbspersonen. In Deutsches Zentrum für Altersfragen (Ed.), *Beschäftigungssituation älterer Arbeitnehmer. Expertisen zum Fünften Altenbericht der Bundesregierung* (pp. 93–128). Berlin: LIT.

Begemann, V. (2009). Bürgerschaftliches Engagement in Wissenschaft und Praxis. Ein strukturiertes Studienangebot für Ältere an der Westfälischen Wilhelms-Universität Münster. In F. Sagebiel (Ed.), *Flügel wachsen. Wissenschaftliche Weiterbildung im Alter zwischen Hochschulreform und demografischem Wandel* (pp. 26–36). Münster: LIT.

Bertram, T., Dabo-Cruz, S., Pauls, K., & Vesper, M. (2017). Bundesarbeitsgemeinschaft Wissenschaftliche Weiterbildung für Ältere (BAG WiWA). In B. Hörr & W. Jütte (Eds.), *Weiterbildung an Hochschulen. Der Beitrag der DGWF zur Förderung wissenschaftlicher Weiterbildung* (pp. 73–84). Bielefeld: wbv.

Brookfield, S. D. (2009). Self-directed learning. In R. Maclean & D. Wilson (Eds.), *International handbook of education for the changing world of work* (pp. 2615–2627). Dordrecht: Springer.

Bundesministerium für Bildung und Forschung. (2015). *Weiterbildungsverhalten in Deutschland 2014. Ergebnisse des Adult Education Survey.* AES Trendbericht. https://www.bmbf.de/pub/Weiterbildungsverhalten_in_Deutschland_2014.pdf. Accessed July 29, 2017.

Costard, A., Haller, M., Meyer-Wolters, H., & Pietsch-Lindt, U. (2012). Alter forscht! Forschendes Lernen, Aktionsforschung und Ageing Studies im Seniorenstudium. In A. Costard, M. Haller, H. Meyer-Wolters, & U. Pietsch-Lindt (Eds.), *Alter forscht! Forschungsaktivitäten im Seniorenstudium. Forschendes Lernen, Aktionsforschung und ageing studies. Jahrestagung 2009* (pp. 5–8). Hamburg: DGWF.

Dabo-Cruz, S., & Pauls, K. (2018). 30 Jahre Seniorenstudium—eine kritische Zwischenbilanz. In R. Schramek, J. Steinfort-Diedenhofen, B. Schmidt-Hertha, & C. Kricheldorff (Eds.), *Alter(n), Lernen, Bildung. Theorien, Konzepte und Diskurse* (pp. 175–186). Stuttgart: Kohlhammer.

Dummer, I. (2007). Bedeutung von Interessensvertretung und Engagement Seniorenstudierender für die Entwicklung wissenschaftlicher Weiterbildungsangebote für Ältere. In M. Kaiser (Ed.), *Studium im Alter—Eine Investition in die Zukunft?!* (pp. 109–117). Münster: Waxmann.

European Federation of Older Students. (2018). *European Federation of Older Students.* http://www.efos-europa.eu/. Accessed June 12, 2018.

Feld, T. C., & Franz, M. (2016). Wissenschaftliche Weiterbildung als Gestaltungsfeld universitären Bildungsmanagements. Ergebnisse einer explorativen Fallstudie. *Zeitschrift für Pädagogik, 62*(4), 513–530.

Field, J., Schmidt-Hertha, B., & Waxenegger, A. (Eds.). (2015). *Universities and engagement: International perspectives.* London/New York: Routledge.

Formosa, M. (2012). Education and older adults at the University of the Third Age. *Educational Gerontology, 38*(2), 114–126.

Formosa, M. (2014). Four decades of Universities of the Third Age: Past, present, and future. *Ageing & Society, 34*(1), 42–66.

Formosa, M. (2016). Malta. In B. Findsen & M. Formosa (Eds.), *International perspectives on older adult education: Research, policies, practices* (pp. 261–272). Cham, Switzerland: Springer.

Gösken, E. (2012). Intergenerationelles Lernen an der TU Dortmund. *Journal Hochschuldidaktik, 23*(1–2), 17–19. https://eldorado.tu-dortmund.de/bitstream/2003/29701/1/journal_hd_1-2_2012_goesken.pdf. Accessed July 29, 2017.

Green, A. (2016). *Research-based learning strategies for successfully linking teaching and research.* http://www.griffith.edu.au/learning-futures/pdf/gihe_tipsheet_web_rbl.pdf. Accessed April 28, 2018.

Hochschulforum Digitalisierung. (2016). *Hochschulbildung im digitalen Zeitalter* [The Digital Turn]. https://hochschulforumdigitalisierung.de/sites/default/files/dateien/Abschlussbericht.pdf. Accessed Jan 15, 2018.

Kaiser, M. (2009). Eine neue Generation älterer Studierender? Ein Blick auf die Teilnehmer und Teilnehmerinnen des "Studium im Alter" an der Westfälischen Wilhelms-Universität Münster. In F. Sagebiel (Ed.), *Flügel wachsen. Wissenschaftliche Weiterbildung im Alter zwischen Hochschulreform und demografischem Wandel* (pp. 93–106). Münster: LIT.

Koch, W., & Frees, B. (2017). ARD/ZDF-Onlinestudie 2017: Neun von zehn Deutschen online. *Media Perspektiven, 2017*(9), 434–446.

Kraft, S. (1999). Selbstgesteuertes lernen. Problembereiche in theorie und praxis. *Zeitschrift für Pädagogik, 45*(6), 833–845.

Kröll, M. (2011). Motivstrukturen zur wissenschaftlichen Weiterbildung. In U. Faßhauer, J. Aff, B. Fürstenau, & E. Wuttke (Eds.), *Lehr-lernforschung und professionalisierung. Perspektiven der Berufsbildungsforschung* (pp. 173–185). Opladen: Budrich.

Ladas, H., & Levermann, U. (2007). Förderung des intergenerationellen Dialogs durch ein Studium im Alter. In M. Kaiser (Ed.), *Studium im Alter—Eine Investition in Zukunft?* (pp. 141–147). Münster: Waxmann.

Low, R., & Jin, P. (2012). Self-organized learning. In N. M. Seel (Ed.), *Encyclopedia of the sciences of learning.* Boston: Springer.

Malwitz-Schütte, M. (2000). Selbstgesteuertes und selbstorganisiertes Lernen in der Weiterbildung älterer Erwachsener—ein Konzept macht Furore. In Deutsches Institut für Erwachsenenbildung (Ed.), *Selbstgesteuerte Lernprozesse älterer Erwachsener. Im Kontext wissenschaftlicher Weiterbildung* (pp. 37–64). Bielefeld: Bertelsmann.

Marotzki, W. (1990). *Entwurf einer strukturalen Bildungstheorie. Biographietheoretische Auslegung von Bildungsprozessen in hochkomplexen Gesellschaften.* Weinheim: Deutscher Studien Verlag.

Martin, A. (2016). Das Weiterbildungspersonal demografisch. In A. Martin, S. Lencer, J. Schrader, S. Koscheck, H. Ohly, R. Dobischat, A. Elias, & A. Rosendahl (Eds.), *Das Personal in der Weiterbildung* (pp. 63–68). Wiesbaden: WBV.

Oldenburger Erklärung. (2018). *Oldenburger Declaration.* https://dgwf.net/arbeitsgemeinschaften/bag-wiwa/aktuelles/article/oldenburger-erklaerung. Accessed April 6, 2018.

Pintrich, P. R., & de Groot, E. V. (1990). Motivational and self-regulated learning components of classroom academic performance. *Journal of Educational Psychology, 82*(1), 33–40.

Sagebiel, F. (2006). Demographischer Wandel und Bolognaprozess. Auswirkungen auf das Seniorenstudium? *Hochschule und Weiterbildung, 2006*(2), 9–19.

Sagebiel, F. (2014). Best-Practice-Ansätze in der allgemeinen wissenschaftlichen Weiterbildung für Ältere. *Hochschule und Weiterbildung, 2014*(1), 41–45.

Sagebiel, F., & Dahmen, J. (2009). Neue Trends im Seniorenstudium. Zwischenergebnisse der BAG WiWA Studie. In F. Sagebiel (Ed.), *Flügel wachsen. Wissenschaftliche Weiterbildung im Alter zwischen Hochschulreform und demografischem Wandel* (pp. 14–25). Münster: LIT.

Saup, W. (2001). *Studienführer für Senioren.* Berlin: Bundesministerium für Bildung und Forschung.

Schmidt, B. (2007). Older employee behaviour and interest in continuing education. *Journal of Adult and Continuing Education, 13*(2), 156–174.

Schmidt-Hertha, B. (2014a). *Kompetenzerwerb und Lernen im Alter. Studientexte für Erwachsenenbildung.* Bielefeld: wbv.

Schmidt-Hertha, B. (2014b). Technologiebasierte Problemlösekompetenz. In J. Friebe, B. Schmidt-Hertha, & R. Tippelt (Eds.), *Kompetenzen im höheren Lebensalter. Ergebnisse der Studie "Competencies in Later Life" (CiLL)* (pp. 99–114). Bielefeld: Bertelsmann.

Schmidt-Hertha, B. (2017). Die Rolle der Hochschulen für die Weiterbildung im dritten Lebensalter. *Zoom, 7,* 11–20.

Schmidt-Hertha, B. (2018). Wissenschaftliche Weiterbildung für Ältere. In W. Jütte, & M. Rohs (Eds.), *Handbuch Wissenschaftliche Weiterbildung.* Bielefeld: Springer (in press).

Schmidt-Hertha, B., Jelenc Krašovec, S., & Formosa, M. (Eds.). (2014). *Learning across generations: Contemporary issues in older adult education.* Rotterdam, Netherlands: Sense Publishers.

Schmidt-Hertha, B., & Muehlbauer, C. (2012). Lebensbedingungen, Lebensstile und Altersbilder älterer Erwachsener. In F. Berner, J. Rossow, & K.-P. Schwitzer (Eds.), *Individuelle und kulturelle Altersbilder. Expertisen zum Sechsten Altenbericht der Bundesregierung. Band 1* (pp. 109–149). Wiesbaden: VS Verlag.

Seidel, E., & Siebert, H. (1998). Seniorenstudium als Konstruktion von Wirklichkeit. In M. Malwitz-Schütte (Ed.), Lernen im Alter. *Wissenschaftliche Weiterbildung für ältere Erwachsene* (pp. 57–76). Münster: Waxmann.

Stadelhofer, C. (1999). Lebenslanges Lernen und Altenbildung im europäischen Kontext. In R. Bergold, D. Knopf, & A. Mörchen (Eds.), *Altersbildung an der Schwelle des neuen Jahrhunderts* (pp. 116–122). Bonn: Echter.

Stadelhofer, C. (2000a). "Forschendes Lernen" im dritten Lebensalter. In S. Becker, L. Veelken, & K. P. Wallraven (Eds.), *Handbuch Altenbildung. Theorien und Konzepte für Gegenwart und Zukunft* (pp. 304–310). Opladen: Leske & Budrich.

Stadelhofer, C. (2000b). Selbstgesteuertes Lernen und Neue Kommunikationstechnologien als neue Wegpfeiler in der Altenbildung. In S. Becker, L. Veelken, & K. P. Wallraven (Eds.), *Handbuch Altenbildung. Theorien und Konzepte für Gegenwart und Zukunft* (pp. 255–267). Opladen: Leske & Budrich.

Stahl, D. (2011, February 9). Die Rentner von der ersten Bank. Zeit online. http://www.zeit.de/studium/uni-leben/2011-02/seniorenstudium-hochschulen. Accessed July 26, 2017.

Statistisches Bundesamt. (2017). *Bildungsstand der Bevölkerung 2016.* Wiesbaden: Destatis.

Steinhoff, B. (1997). Die Beziehungen von Jung und Alt im Studium—Empirische Befunde und Fragen. *Hessische Blätter für Volksbildung, 47*(2), 136–153.

Swindell, R., & Thompson, J. (1995). An international perspective on the university of the third age. *Educational Gerontology, 21*(5), 429–447.

Tippelt, R., Schmidt, B., Schnurr, S., Sinner, S., & Theisen, C. (Eds.). (2009). *Bildung Älterer—Chancen im demografischen Wandel.* Bielefeld: Bertelsmann.

Tremmel, J., Bschaden, A., & Wagermaier, A. (2014). Die wachsende Zahl von Seniorenstudierenden in Deutschland unter dem Aspekt der Generationengerechtigkeit: Harmonie oder Konflikt? In J. Tremmel (Ed.), *Generationengerechte und nachhaltige Bildungspolitik* (pp. 203–241). Wiesbaden: Springer VS.

Tremp, P. (2015). Forschungsorientierung und Berufsbezug. Notwendige Relationierungen in Hochschulstudiengängen. In P. Tremp (Ed.), *Forschungsorientierung und Berufsbezug im Studium. Hochschulen als Orte der Wissensgenerierung und der Vorstrukturierung von Berufstätigkeit* (pp. 13–39). Bielefeld: Bertelsmann.

Dr. Bernhard Schmidt-Hertha is Professor of Educational Research with a focus on vocational continuing education and on-the-job training at the University of Tuebingen (Germany). He studied educational research, psychology and sociology in Munich, where he finished his Ph.D. in

2004 and his habilitation in 2009. Dr. Schmidt-Hertha is co-editor of an online-journal, reviewer for the German Research Association, and different national and international journals. Since 2010, he is convener of the ESREA Network on Education and Learning of Older Adults (ELOA). Currently, Dr. Schmidt-Hertha is working on digitalisation in adult education, education and learning in later life, and the experience of early school leavers in adult education. He is also interested in research on quality in higher education.

Chapter 7
Third Age Learning for Active Ageing in Malta: Successes and Limitations

Marvin Formosa

Introduction

A lot of water has gone under the bridge since the launching of a University of the Third Age (U3A) in Malta in 1993. The Maltese U3A runs from October to June and operates from five centres—namely, in Floriana, Sliema, Attard, Vittoriosa, and Għajnsielem (Gozo). Whilst the premises in Floriana are spacious and can hold up to 200 members, the other premises are relatively smaller, with a maximum capacity of about 60 persons. In June 2018, the Maltese U3A included 654 members, 179 men and 492 women—thus, female members outnumbered males by a ratio of three to one. The majority of members tend to be in the 60–69 age cohort (42%), with both membership and participation falling with increasing age: 70–79 (39%), 80–89 (11%), and 90-plus (2%). Six per cent declined to list their exact ages. An account of the historical development of the U3A in Malta is not the objective of this chapter as it can be found in much detail elsewhere (Formosa, 2000, 2002, 2005, 2007, 2012, 2016). Instead, the aim of this chapter is to report on a multi-method study investigating the impact on third age learning on active ageing. Such a study was warranted because whilst many research publications applaud U3As for their potential to bring about higher levels of well-being amongst older persons, empirical research documenting the potential successes and limitations of U3As in bringing about higher levels of active ageing amongst its members is relatively lacking.

M. Formosa (✉)
University of Malta, Msida, Malta
e-mail: marvin.formosa@um.edu.mt

© Springer Nature Switzerland AG 2019
M. Formosa (ed.), *The University of the Third Age and Active Ageing*, International Perspectives on Aging 23, https://doi.org/10.1007/978-3-030-21515-6_7

The Operationalisation of Active Ageing

Following the World Health Organization's (WHO) (2002: 12) definition of active ageing as 'the process of optimizing opportunities for health, participation and security in order to enhance quality of life as people age', one witnessed a number of efforts to construct an empirical scale that measures the extent that older persons are experiencing an active ageing lifestyle. For instance, Bowling (2008) conducted face-to-face structured interviews with 337 people aged 65-plus living at home in Britain and found that the most common perceptions of active ageing were 'having/maintaining physical health and functioning' (43%), 'leisure and social activities' (34%), 'mental functioning and activity' (18%), and 'social relationships and contacts' (15%). She concluded that the

> People's views focussed on basic definitions such as social, physical and mental health and activity, probably reflecting the novelty of the concept to them, thereby excluding frail older people from active ageing. Comparisons with definitions of successful ageing and quality of life showed overlap, but the latter were portrayed as 'states of being'. This is consistent with models which propose quality of life as the end-point of active ageing.
>
> Bowling, 2008: 293

Presently, one popular empirical construct that ranks countries according to different indicators of active ageing, with a wide application in both Europe and elsewhere, is the Active Ageing Index (AAI). The AAI developed in the context of the 2012 European Year for Active Ageing and Solidarity between Generations and aimed at raising awareness of population ageing and the positive solutions towards the challenges it brings (Zaidi, Harper, & Howse, 2018). Reflecting the multidimensional concept of active ageing, the AAI is hinged upon four domains—namely, 'employment', 'participation in society', 'independent, healthy, and secure living', and 'capacity and enabling environment'. Whilst the former three domains refer to the actual experiences of active ageing (employment, unpaid work/social participation, and independent living), the fourth domain captures the capacity for active ageing as determined by individual characteristics and environmental factors. Whilst emergent research recognised the contribution of the AAI in sensitising policymakers to the multidimensionality and complexity of the process of 'ageing well' (Formosa, 2017), critics also noted that the AAI remains under-scrutinised and under-theorised:

> This model is expert-based and ingrained with a priori assumptions about the potential of older people, the domains of life and activities they value and how strongly they value them…the Active Ageing Index measures current achievements, not capabilities (i.e. the opportunity set of achievable "doings" and "beings"), resulting in a valuable but incomplete tool for policymaking purposes.
>
> de São José, Timonen, Alexandra, Amado, & Pereira Santos, 2017: 49

In another attempt at operationalising active ageing, Paúl, Ribeiro, and Teixeira (2012) published his data from a sample of 1,322 community-dwelling older persons, and through confirmatory factor analysis found that the emergent results failed to confirm the WHO conceptual model. This was because some of the determinants for active ageing were highly intertwined, and psychological factors were found to play a larger part than the level accredited by the WHO. Alternatively, they forwarded an alternative model that accounted for more than 50% of the data and which included six factors—namely, health, psychological components, cognitive performance, social relationships, bio-behavioural components, and personality. Although Paúl, Ribeiro, and Teixeira's study demonstrated that whilst there are both objective and subjective variables contributing to active ageing, and that psychological variables seem to be key determinants to the construct, their analysis was criticised for a lack of clarity and for proposing a 'measurement model based on the determinants of active ageing rather than on indicators of the concept' (Bélanger, Ahmed, Filiatrault, Yu, & Zunzunegui, 2017: 198). Elsewhere, Fernández-Ballesteros, Robine, Walker, and Kalache (2013: 2) maintained that although there is no empirical definition of active ageing that is universally accepted, there is a consensus that it includes the following domains: 'low probability of illness and disability, high physical fitness, high cognitive functioning, positive mood and coping with stress, and being engaged with life'. More recently, Bélanger et al. (2017) compared the WHO's and Fernández-Ballesteros et al., (2013) models of active ageing with epidemiological data from Canada through data which included 799 community-dwelling older adults between 65 and 74 years old, recruited from the patient lists of family physicians in Saint-Hyacinthe, Quebec and Kingston, in Ontario. The study concluded that

> Neither could be validated in the sample of Canadian older adults. Although a concept of healthy aging can be modeled adequately, social participation and security did not fit a latent factor model. A simple binary index indicated that 27% of older adults in the sample did not meet the active aging criteria proposed by the WHO.
>
> Bélanger et al., 2017: 197

Bélanger et al. (2017: 197) advocated that active ageing might represent a 'human rights policy orientation rather than an empirical measurement tool to guide research among older adult populations' and that 'binary indexes of active aging may serve to highlight what remains to be improved about the health, participation, and security of growing populations of older adults'. Another scale seeking to measure active ageing constitutes the University of Jyvaskyla Active Ageing Scale (Rantanen et al., 2018), which consisted of a 17-item scale whereby each item is researched in terms of the respondents' 'will to act', 'ability to act', 'possibility to act', and 'frequency of doing'. In comparison with previous empirical models, the strength of the Jyvaskyla Active Ageing Scale included the 'definition of active aging at the level of the individual that was used as the foundation for its development, the novelty of developing a scale for assessing active aging as a quantifiable construct, the item response analyses, and the participant involvement' (ibid., 2018: 21).

The WHO's approach towards active ageing has also questionable as it includes a degree of cultural bias, since its knowledge base is overwhelmingly based on Western studies, in that its discourses tend to be deeply 'embedded within the U.S. concept of productive aging…and strongly U.S. culture-bound' (Walker, 2009: 86). For example, focusing on the Thai cultural context, Thanakwang, Isaramalai, and

Hatthakit's (2014) study generated a six-themed scale for measuring active ageing in Thailand—namely, self-reliance, social participation, spirituality, healthy living, active learning, and income security management. The authors concluded that although the 'perceptions of active ageing among the Thai elderly [sic] involved health, social participation, and security in life, which are also the three key pillars of active ageing suggested by WHO', when 'compared to research in a Western context, some of the dimensions of Thai active ageing were distinct, specifically growing spirituality and managing later life security, while others were overlapping' (ibid.: 152). There is no doubt that in so much as Asian older persons tend to perceive active ageing as comprising healthy living, social engagement, and security—for example, Hong Kong Chinese older persons viewed active ageing as comprising good health, positive life attitude, active social engagement, possessing social and financial capital, and living in meaningful residences—as well as the criticism levelled at positivist constructed scales of measurements, there is a real urgency to commence working on bottom-up and culturally sensitive capacities of active ageing.

Research Methodology

This chapter reports on the carrying out of a multi-method study to research the impact of participation in the Maltese U3A on bringing about and/or improving active ageing lifestyles in the second part of the life course. As a multi-method research design 'is the conduct of two or more research methods, each conducted rigorously, and complete in itself, in one project' (Morse, 2003: 190), this study opted for a sequential multiphase QUAL-QUAL study in which 'more than two phases or both sequential and concurrent strands are combined over a period of time within a program of study addressing an overall program objective' (Schoonenboom & Burke Johnson, 2017: 118). The *first* research phase consisted of carrying out exploratory research to resolve the tension between objective definitions and subjective understandings of active ageing amongst older Maltese learners. In July 2018, two focus groups were held with U3A members, selected through purposive sampling, and each including nine members, five women and four men. Following an ice-breaker, the author presented—in a pictographic manner—the WHO's definition of active ageing and the different measurements of active ageing reviewed in this chapter's second section to focus group participants. Following an hour-long discussion in each different focus group, data was subsequently analysed through 'logical analysis' which revealed the logical shape of informants' ideas by first locating premises within the data that symbolised one group and then exploring connections between one group of premises and another (Miles & Huberman, 1994). As elaborated upon in the next section, data analyses uncovered that older learners at the U3A understood and experienced active ageing as hinging on three platforms—namely, health, social inclusion/participation, and independence. The *second* research phase consisted of conducting semi-structured interviews with older learners attending the Maltese U3A. This time around, interviewees were selected through convenience sampling and were asked questions based on the emergent data analysis during first phase of the multi-method study (Box 7.1).

Box 7.1: Second Research Phase: Key Interview Prompts
Health

- How is your health?
- To what extent do you perceive the U3A as impacting your health?
- Do you perceive such impacts as positive or negative?

Social inclusion and participation

- Do you feel part of the Maltese social community?
- Do you feel as belonging to a specific group of friends or acquaintances?
- To what extent do you perceive the U3A as impacting your participation in society?
- Do you perceive such impacts on your social participation in the community as positive or negative?

Independence

- Do you consider yourself as an independent person?
- Are you able to achieve your goals and objectives in daily life?
- To what extent do you perceive the U3A as impacting your goal to be independent in your daily life?
- Do you perceive such impacts as positive or negative effects?

Data analyses followed open, in vivo. and selective coding strategies. Initial coding consists in breaking down qualitative data into discrete parts, closely examining them, and comparing them for similarities and differences (Strauss & Corbin, 1998) to remain open to all possible theoretical directions indicated by one's reading of the data (Charmaz, 2006). In vivo coding enables researchers to 'preserve participants' meanings of their views and actions in the coding itself' (ibid.: 55)—thus, providing imagery, symbols, and metaphors for rich category, theme, and concept development. Herein, researchers thus look for words or phrases that seem to stand out, for example, nouns with impact, action-orientated verbs, evocative word choices, clever phrases, or metaphors. Finally, selective coding functions like an umbrella that covers and accounts for all other codes, the core category which consists of all the products of analysis condensed into a few words that seem to explain what 'this research is all about' (Strauss & Corbin, 1998: 146).

Older Learners' Experience and Understanding of Active Ageing

Three themes emerged from the first phase of the study with regard to the subjective perceptions of older learners as to what constitutes the key pillars of active ageing—namely, health, social inclusion and participation, and independence. There was a consensus in both focus groups that 'health' is the most important determinant for active ageing, as it facilitated the possibility to participate in both solitary and social activities. As a counterpoint, 'pain' was perceived as being a key obstacle towards leading active and successful ageing lifestyles, on the basis that it hinders people from both a serene and vigorous life:

> To be active, one needs to be healthy, to be able to wake up in the morning and face the day without any pain. If one is ill, then one tends to focus all his/her energies on getting back in good shape. When I am ill, the world stops. Not only am I not able to live an active lifestyle, but I close into myself dreading the day during which I cannot do anything, and dreading the night during which I cannot sleep. (focus group participant, male)

Older learners also considered ill-health and pain to have a drastic negative effect on one's psychological well-being, and many recounted their personal experiences or those of relatives or friends, who experienced depression and other mental health issues due to chronic pain.

> My friend was unrecognisable when she was ill. She became ill-tempered, morose, and pessimistic about everything. Sometimes I phoned her and she did not even pick up the phone. One's interest in activities dissipates as soon as your mental health suffers a dip. Even with bad news, I find myself withdrawing from my family and friends… Depression is very common when one is frail. Depression kills you socially! (focus group participant, female)

Pain was envisaged by focus group members to impact negatively on older persons' self-esteem and self-confidence which, in turn, rendered sufferers homebound due to no interest or willingness to participate in social events. The ability for social participation was considered as another *sine qua non* for the possibility of active ageing in later life, in that focus group members singled out 'social exclusion' as the key determinant for loneliness and solitude in later life:

> To be an active ager requires one to be in the thick of things. Can you be active but alone? Perhaps for a short period but not for long. Active ageing is all about meeting people, meeting your friends, befriending new ones, and participating in social events…Social events are the highlight of an active ageing lifestyle. It is the time one shares one's life with others. Social events are the epitome of active ageing. (focus group participant, male)

All members agreed that traditional familial dynamics as such are now something of the past. Whilst just until a quarter of a century ago, daughters generally lived in the same city or village as their parents, so that there was both frequent and warm contact between different familial generations, in current times this is more the exception than the rule. As a result, younger family members cannot be dependent upon as far as social participation is concerned as much as they were in the past, and today older persons have to draw on friends, acquaintances and non-governmental organisations as sources of social bonding, inclusion, and informal care:

> The family, as we know when younger, has all but disappeared. My grandchildren are like foreigners to me. I have not seen them for weeks now! My daughter visits but she is always in a hurry. I do not blame her, she works as a [professional occupation], it is not easy to balance family and work responsibilities…I thank God that I have friends to go out with. The Local Authority is also very active as far as social outings are concerned. (focus group participant, female)

The risk of experiencing social exclusion was a current topic during the focus groups, especially for unmarried and widowed persons. This was not unexpected since the 'double jeopardy' experience, whereby two or more characteristics combine to create a 'double disadvantage', is a key motif in gerontological research. For example, local studies indicate a strong prevalence of poverty amongst older women compared to men (Formosa, 2017). Moreover, older women are more likely to be victims of crime and hold lower levels of education, and because many tend to spend their final stage of life as widows, they tend to enter long-term care due to a lack of available care-givers (ibid.). Finally, as Thanakwang, Isaramalai, and Hatthakit (2014) found out in their study with Thai citizens, being independent was viewed as another key aspect of active ageing and a process that facilitates older persons' self-reliance—thus, assuring that they are not a burden to their families or reliant on either informal or formal carers. Focus members emphasised the need to be able to count on themselves without banking on others, and hence, feel free to execute all wishes and desires as independently as possible:

> One cannot be active if one cannot decide here and then what one wishes to do during the day. We are all taking part in this discussion because we accepted your invite and decided to transport ourselves to this room. I would not have been able to come if I had to depend on my children to get me here. They might have had other things to do. (focus group participant, female)

Older learners ranked very high the need to be able to live independently, without any form of support, whilst also giving importance to autonomous decision-making. They all loathed the possibility that one day they would have to ask others to do things for them, and extreme situations, to bathe and dress them:

> Active ageing is all about being in charge. One's you lose your mobility, and God forbid, your mental faculties, you are doomed. How can one be active when one needs others to provide the required assistance even permission? One must work hard to remain as independent as long as possible, even if with the help of assistive technology. (focus group participant, male)

Participants indicated that being able to take all the required decisions meant that they were self-reliant, and hence, could decide on both their type and level of active ageing lifestyle:

> I constantly hear that ageing is also a time for new opportunities, to do all those things that one wished to do but never had the time. True? Yes, but only to an extent! One needs to be healthy to be able to perform desired activities, such as going abroad … When people are deciding for you, even if these people are your children, and they mean well, life's opportunities decrease substantially. (focus group participant, female)

Being independent was perceived as a vehicle to participate in meaningful activities, and to provide them with the possibility to engage in the community in a productive manner, both in terms of taking part in municipal events and to contribute to the community's social capital.

'Active Ageing' Through 'Third Age Learning'

The second phase of the study elicited the potential impact that participation in the Maltese U3A enables participants to reach higher levels of active ageing lifestyles. During interviews, respondents acknowledged that they have no hard evidence that participating in the Maltese U3A in Malta led to an improvement in their physical health, but they were adamant that when—due to circumstances beyond their control—they don't attend lectures, they feel less satisfied with their physical well-being. In one interviewee's words,

> Attending the U3A makes me feel well, gives me energy, helping me to forget my aching body. My body relaxes. As I learn, as I meet my friends, as I laugh, I fell re-born, I feel young, strong and healthy. I cannot explain what I feel. I guess everything is in the mind, but the effects are real. I walk to the U3A centre with a spring in my step, walking swiftly, and effortlessly. The U3A really effects my health positively! (interviewee, male)

Another interviewee held a pragmatic outlook on the health benefits of attending the U3A. He was of the opinion that the quest to attend the learning sessions had a great impact on his daily choices which, in turn, made him healthier and more physically robust:

> The U3A keeps you active and this has a lot of positive implications health-wise. As I live far away from the U3A Centre, I wake up early, shower, make an effort to dress well, I groom myself, and eat a healthy breakfast. Just by attending, I have to walk for five minutes to the bus stop and then another five minutes to the centre. After the learning programme finishes I always join my friends for a walk. If it were not for the U3A, I would be living a very sedentary life. (interviewee, male)

All interviewees emphasised how participating in the U3A permeated them with higher levels of self-esteem, self-satisfaction, and happiness. The U3A had profound meaning for participants, especially women who were formally employed in the workforce—as finally they felt that they have an opportunity to succeed and contribute outside a familial environment. Many felt reborn, and highly motivated and willing to take on volunteer administrative roles within the U3A. For some female members, such benefits came as a surprise to them:

> I joined the U3A only because my friend joined and she did not wish to attend the learning programmes on her own. It was one of the best decisions of my life. I am not exaggerating. After years of toiling in house chores I am now living a new life. I still cook for my husband but now mealtimes have to accommodate my U3A timings. I never thought later life would be such fun. (interviewee, female)

> I enrolled in the U3A because all my circle of friends joined, and I followed suit. I expected a patronising and snobbish environment, but I could not have been more mistaken. It is a very friendly environment and the opportunities for improving oneself are endless. You can lecture if you want or join the administrative committee or help during the monthly social event. It is a real family here. (interviewee, female)

The participation in the U3A influenced the formation of positive self-images for participants that mitigated the dominant ageist images that one is bombarded with in the social and mass media. It is also noteworthy that participants voiced much support towards the association between learning and keeping cognitively alert. Many claimed that listening attentively to the lecturers and facilitators improved their memory and 'jogged their brain'. At the same time, it was clear that the Maltese U3A was instrumental to improving its members' opportunities for social inclusion and participation. Contrary to existing literature, which highlighted only the role of U3As in enabling older women, especially widows, to reach out and befriend same-aged peers (Robbins-Ruszkowski, 2017), from the data, it resulted that older men were equal beneficiaries. Whilst women tended to make use of their membership in the Maltese U3A to extend their social network, men claimed to relish their participation to mitigate against the feeling of a 'roleless role' following retirement from the labour market:

> After 43 years working in the public service, the first few weeks of retirement, a *forced* retirement to be precise, were hell. Of course you can read, take walks, go abroad. But such activities did not work for me. I need to be part of a team. I do not enjoy solitary activities. The U3A is more than a team, with some 600 members there is always an extensive 'to do' list administration-wise. I feel alive now. (interviewee, male)

It was clear that participation in the U3A also imbued participants with higher levels of social capital, albeit of a 'bonding' rather than a 'bridging' type. Whilst the former refers to social relations with peers with similar socio-economic backgrounds, the latter denotes acquaintances with peers of either lower or higher social standing:

> Before I joined the U3A I used to spend whole days indoors. The weekends were the worst. Having no one to spend one's free time with was very depressing. Here I met many old acquaintances of mine, some of whom were with me at school, whilst also making new friends. I enjoy meeting them because we all have similar backgrounds, and comparable interests and opinions. Sometimes we plan Sunday morning visits to Valletta or afternoon walks. I still spend some days alone but never the whole weekend now. (interviewee, male)

Finally, the data also uncovered some degree of association between attending the Maltese U3A and independence. Such an association revolved largely around the fact that members disclosed that peers at the U3A were always ready to assist in times of duress, and that many learning programmes—ranging from health literacy to legal/notarial to social work curricula—were extremely helpful in providing them with the required knowledge to remain as independent as possible in later life. Whilst such assistance seems, *prima facie*, an antidote to achieving independence, interviewees accentuated that help was always forthcoming in an empowering manner rather than in a patronising one:

I am very grateful to the U3A in a way that perhaps you may not understand. I was going through a bad time last year, when I had some family problems, but coming here gave me not only the required psychological energy to face the daily challenges, but also expert advice and counsel. My friends here were very patient with me and tried to help me help myself. They never treated me like a child but as a fellow friend…The study programmes on legal and notarial affairs were also a great help, and some lecturers gave me complimentary expert advice during the coffee break. (interviewee, female)

One area which many U3A members believed was highly beneficial to their potential for independence was the organisation's availability of learning programmes on information and communication technology. Such curricula reflected the increasing requirement for citizens to be digitally literate and competent, and included learning word processing, electronic mail, Internet surfing, online banking, online searching, password safety, and accessing online information on public welfare and pension services. Indeed, an engagement in digital citizenship in contemporary societies is an important factor to the full social inclusion of citizens, particularly as more public services go online (Formosa, 2013). In one interviewee's words, 'learning the internet and how to use my tablet gave me the independence I craved for…no longer waiting for others to help you book flights on the internet, or waiting for my children to visit me. Now, I am also booking my hairdresser by email' (interviewee, female).

Conclusion

There is no doubt that the U3A in Malta is a key vehicle for bringing about and improving active ageing lifestyles amongst older persons. The U3A offered a promising range of study programmes ranging from purely academic pursuits to more artistic ones to other events of a more social nature. It provided learners with both humanistic proficiencies such as when learning new languages to more vocational skills such as when learning online communications through software such as Skype and WhatsApp. The Maltese U3A is therefore to be credited as being at the forefront of attempts to transform later life from one of social stigma and negative labelling to a period of positive ageing by improving physical, emotional, and social well-being. The Maltese U3A is typified by a sense of vitality and dynamism that goes beyond what is usually the case in normal educational classes. It fulfils various positive social and personal functions such as aiding lonely older persons to re-socialise themselves, as well as providing opportunities and stimulation for the use and structure of free-time which would otherwise be characterised by inactivity. As argued elsewhere, participation at the Maltese U3A developed

…among learners a lofty and progressive delight of life, increased the social integration and harmony of older persons in society, and injected them with a sense of creativity, whilst making them more visible in society. It…improved members' abilities of understanding the objective world by aiding them to grasp better global development and social progress, and helped them to ameliorate their abilities of self-health by enabling them to master medical care knowledge and prevention of disease.

Formosa, 2012: 283

However, this study also exposed a number of limitations as far as active ageing is concerned, since participation in the Maltese U3A was also exemplified by a number of exclusionary forces that question its suitability as a wide and equitable 'active ageing' *modus operandi*. Data pointed clearly to the fact that the benefits of the U3A are being mostly taken up by women, older persons of higher-than-average socio-economic status, and older persons who enjoy good physical and cognitive health status. First, as in Poland the 'gendered culturally ideals of sociability over the life course shape participation in U3As...as gender segregation at the U3A connects to broader discourses of ageing and gender' (Robbins-Ruszkowski, 2017: 117). Second, the U3A has clearly more appeal to older persons from career-oriented backgrounds which required higher-than-average levels of educational backgrounds. Older persons whose life experience included manual labour, the service industries and technical expertise were in short supply if present at all. The predominance of a middle-class ethos at the U3A was not a coincidence and the result of a number of social closure tactics which made the learning experience unappealing to older persons with low levels of income and education. The choice of both academic and artistic subjects provided an alien environment to older persons from working-class milieus. Unfortunately, the U3A did not escape the pervasiveness of schooling as its organisation operated through a top-down model of instruction that cultivates respect for authority, experts, and universal knowledge. Of course, the term 'university' in its title does not help as working-class persons tend to be apprehensive to join an organisation with such a heavy class baggage, and one witnessed no efforts on behalf of the entity to smooth such barriers. Finally, there seems to be an unwritten but entrenched rule that the U3A, and active ageing as such, is prerogative of only those older persons with good bodily health. No older persons in wheelchairs or using canes were visible and all members were community-dwelling older persons. As a result, older persons with physical mobility difficulties and cognitive challenges were completely absent. In this way, one could argue that the third objective of the U3A, that of improving independence, is a clear case of self-fulfilling prophecy (Formosa, 2014). Whilst this is not the same as saying that the Maltese U3A is an undeserving institution, one has to concur with Robbins-Ruszkowski (ibid.: 119) when she concluded that 'taken together, the gendered, classed, and bodily nature of both the form and content of activities... render these institutions exclusionary...the vision of active ageing that they promote...make it inaccessible to some segment of the population'. With respect to future and required research, such observations warrant urgent 'action research' in third age learning that explores how U3As can embrace an understanding of active and successful ageing that is less hinged and divided along gender, class, and physical capital lines.

References

Bélanger, E., Ahmed, T., Filiatrault, J., Yu, H. T., & Zunzunegui, M. V. (2017). An empirical comparison of different models of active aging in Canada: The international mobility in aging study. *The Gerontologist, 57*(2), 197–205.

Bowling, A. (2008). Enhancing later life: How older people perceive active ageing? *Aging and Mental Health, 12*(3), 293–301.

Charmaz, K. (2006). *Constructing grounded theory: A practical guide through qualitative analysis.* Thousand Oaks, CA: Sage.

de São José, J. M., Timonen, V., Alexandra, C., Amado, F., & Pereira Santos, S. (2017). A critique of the Active Ageing Index. *Journal of Aging Studies, 40*(January), 49–56.

Fernández-Ballesteros, R., Robine, J. M., Walker, A., & Kalache, A. (2013). Active aging: A global goal. *Current Gerontology & Geriatrics Research, 2013,* Article ID 298012.

Formosa, M. (2000). Older adult education in a Maltese University of the Third Age: A critical perspective. *Education and Ageing, 15*(3), 315–339.

Formosa, M. (2002). Critical gerogogy: Developing practical possibilities for critical educational gerontology. *Education and Ageing, 17*(3), 73–86.

Formosa, M. (2005). Feminism and critical educational gerontology: An agenda for good practice. *Ageing International, 30*(4), 396–411.

Formosa, M. (2007). A Bourdieusian interpretation of the University of the Third Age in Malta. *Journal of Maltese Education Research, 4*(2), 1–16.

Formosa, M. (2012). Education and older adults at the University of the Third Age. *Educational Gerontology, 38*(1), 114–125.

Formosa, M. (2013). Digital exclusion in later life: A Maltese case-study. *Humanities and Social Sciences, 1*(1), 21–27.

Formosa, M. (2014). Four decades of Universities of the Third Age: Past, present, and future. *Ageing & Society, 34*(1), 42–66.

Formosa, M. (2016). Malta. In B. Findsen & M. Formosa (Eds.), *International perspectives on older adult education: Research, policies, practices* (pp. 161–272). Cham: Springer.

Formosa, M. (2017). Responding to the Active Ageing Index: Innovations in active ageing policies in Malta. *Journal of Population Ageing, 10*(1), 87–99.

Miles, M. B., & Huberman, A. M. (1994). *Qualitative data analysis.* Thousand Oakes, CA: Sage.

Morse, J. M. (2003). Principles of mixed methods and multi-method research design. In C. Teddlie & A. Tashakkori (Eds.), *Handbook of mixed methods in social and behavioral research* (pp. 189–208). Thousand Oaks, CA: Sage Publication.

Paúl, C., Ribeiro, O., & Teixeira, L. (2012). Active ageing: An empirical approach to the WHO model. *Current Gerontology & Geriatric Research, 2012,* Article ID 382972.

Rantanen, T., Portegijs, E., Kokko, K., Rantakokko, M., Törmäkangas, T., & Saajanaho, M. (2018). Developing an assessment method of active aging: University of Jyväskylä Active Aging Scale. *Journal of Aging and Health.* https://doi.org/10.1177/0898264317750449. Accessed November 30, 2018.

Robbins-Ruszkowski, J. (2017). Aspiring to activity: Universities of the Third Age, gardening, and other forms of living in postsocialist Poland. In S. Lamb (Ed.), *Successful ageing as a contemporary obsession: Global perspectives* (pp. 112–125). New Brunswick, NJ: Rutgers.

Schoonenboom, J., & Burke Johnson, R. (2017). How to construct a mixed methods research design. *Kölner Zeitschrift für Soziologie und Sozialpsychologie, 69*(Supplement 2), 107–131.

Strauss, A., & Corbin, J. (1998). *Basics of qualitative research: Techniques and procedures for developing grounded theory* (2nd ed.). Thousand Oaks, CA: Sage.

Thanakwang, K., Isaramalai, S. A., & Hatthakit, U. (2014). Development and psychometric testing of the active aging scale for Thai adults. *Clinical Interventions in Aging, 9,* 1211–1221.

Walker, A. (2009). Commentary: The emergence and application of active aging in Europe. *Journal of Aging Social Policy, 21*(1), 75–93.

World Health Organization. (2002). *Active ageing: A policy framework*. Geneva: World Health Organization.
Zaidi, A., Harper, S., & Howse, K. (Eds.). (2018). *Building evidence for active ageing policies: Active Ageing Index and its potential*. Singapore: Palgrave Macmillan.

Dr. Marvin Formosa is Associate Professor of Gerontology at the University of Malta where he is Head of the Department of Gerontology and Dementia Studies, Faculty for Social Wellbeing, and contributes to teaching on active ageing, transformative ageing policy, and educational gerontology. He holds the posts of Chairperson of the National Commission for Active Ageing (Malta), Rector's Delegate for the University of the Third Age (Malta), and Director of the International Institute on Ageing, United Nations, Malta (INIA). He directed a number of international training programmes in gerontology, geriatrics and dementia care in the Philippines, China, India, Turkey, Malaysia, Belarus, Kenya, Argentina, Azerbaijan and the Russian Federation. He has published extensively across a range of interests, most notably on active ageing, critical gerontology, Universities of the Third Age and older adult learning. Recent publications included *International perspectives on older adult education* (with Brian Findsen, 2016), *Population ageing in Turkey* (with Yeşim Gökçe Kutsal, 2017), and *Active and healthy ageing: Gerontological and geriatric inquiries* (2018). He holds the posts of Editor-in-Chief of the International Journal on Ageing in Developing Countries, Country Team Leader (Malta) of the Survey of Health, Ageing, and Retirement in Europe (SHARE), and President of the Maltese Association of Gerontology and Geriatrics.

Chapter 8
Late-Life Learning for Social Inclusion: Universities of the Third Age in Poland

Jolanta Maćkowicz and Joanna Wnęk-Gozdek

Historical Overview

The first University of the Third Age (U3A) in Poland was founded on 12 November 1975 in Warsaw by Halina Schwarc (1923–2002) and premised at the Centre of the Third Age. Herein, it is noteworthy to recall key aspects of the founder's life history. In 1939, after Germany and Soviet Russia invaded Poland, she joined the resistance movement at only 16 years of age. After the war, in 1948, she graduated from the Medical University in Poznań and immediately commenced to conduct research on the impact of physical exercise on older people. In 1974, she took her interest in gerontology a step further by setting up an Institute of Gerontology in collaboration with the Department of Gerontological Rehabilitation within the Centre of Post-graduate Medical Education in Warsaw. At this Institute, Schwarc formed a team of medical doctors, psychologists, sociologists, and anthropologists in order to study the issues of ageing, later life, and older persons, from multidisciplinary perspectives. In 1984, she received the title of professor (Polskie Towarzystwo Gerontologiczne, 2017), and in April 1975, during a seminar in Milan dedicated to social welfare met Pierre Vellas who inspired her to found a U3A in Poland. As a result, on 12 November 1975, with the support of the Centre of Postgraduate Medical Education, she succeeded in founding the first Polish U3A at the Centre of the Third Age (Bielowska, 2015). Poland was, as a result, the third ever country, after France and Canada, to boast an educational institution for older and senior citizens (Półturzycki, 2013). The key objective of the U3A included the inclusion of older people into the lifelong education system, as well as to conduct research in the area of social and educational gerontology. Research carried out under Schwarc's leadership consisted in the investigation and analysis of older persons' quality of life in Poland compared with the U3A's membership body. On 4 June 1980, at a conference organised

J. Maćkowicz (✉) · J. Wnęk-Gozdek
Pedagogical University of Cracow, Kraków, Poland
e-mail: jolanta.mackowicz@up.krakow.pl

© Springer Nature Switzerland AG 2019
M. Formosa (ed.), *The University of the Third Age and Active Ageing*, International Perspectives on Aging 23, https://doi.org/10.1007/978-3-030-21515-6_8

by the Centre of the Third Age, it was decided that all associations which work in favour of improving the quality of life and well-being of older persons—through inclusion in the Polish lifelong education system—will be awarded the title of U3A. The dynamics of growth of the U3As between 1976 and 2015 were nothing short of dramatic. Whilst until 1989, there were as little as nine U3As in Poland, in 2007 the number increased to 125, and even doubled by 2010. In 2012, there were almost 400 universities in Poland, and in 2015, over 500 (Polish Association of the Universities of the Third Age, n.d.).

Ageing Policy in Poland

In 2011, the following committees were founded in the Polish Parliament: 'Parliamentary Group for Older People' and 'Parliamentary Group for Universities of the Third Age'. These committees are currently responsible for strengthening the collaboration between those environments in which U3As are generally created—namely, central and regional authorities and academic institutions. In the same year, the Polish Commissioner for Human Rights appointed a 'Commission of Experts for Older People'. Moreover, after the European Commission announced 2012 as the 'European Year of Active Ageing and Solidarity between Generations'; the Polish Senate announced that this year will also represent the 'Year of the Universities of the Third Age', recognising the profound and important role of these institutions. Data from the Polish Association of Universities of the Third Age indicated that the scale of interest in participation in U3As among older persons; according to their database, in October 2015 there were 555 U3As in Poland with over 160,000 members (Szarota, 2015). In August 2012, the Prime Minister's Office was instrumental in establishing a Department of Senior Policy at the Ministry of Labour and Social Policy; in February 2013, the appointment of a Senior Policy Council followed suit (Klimczuk, 2013). Due to the ageing of Polish society, the learning activities of seniors (including U3As) became an integral interest of this Council. The general objective of ageing policy in the area of educational (but also social and cultural) activity is to support the idea of learning in later life and to encourage older persons to engage in civic and social activities. Such policy work has included the development of educational opportunities for older people, the development of leisure programmes for older persons (in particular, civil and voluntary), and an increase in the levels of participation of older people in cultural events, both as creators and spectators.

In recent years, the Polish government adopted the following programmes in the area of ageing policy: *Government Programmes for Social Participation of Senior Citizens* (for 2012–2013 and 2014–2020) as complementary documents to the *Long-Term Senior Policy in Poland* (2014–2020), and *The Perspective of Lifelong Learning*. Whilst the first two documents focus on activities in four priority areas where older adult learning is the main objective, the strategy *The Perspective of Lifelong Learning* (Ministerstwo Edukacji Narodowej, 2013) directs towards facilitating formal, non-formal, and informal education at every stage of life, including later life. It emphasises the importance of learning among older persons, even those aged 80-plus, for both middle- and working-class people, in order to maintain and even improve their key competencies in the Activities of Daily Living. *The Perspective of Lifelong Learning* also addresses the issue of information and communication technology with the objective of increasing the levels of digital competency enjoyed by older Polish persons. Polish senior policy is therefore based on the belief that providing seniors with an opportunity to engage in learning is the main factor determining the upkeep of their social participation in life, and as a result, keeping positive levels of quality of life and well-being well into old age (Ministerstwo pracy i Polityki Społecznej, 2014).

Functioning of Polish U3As

Activities of U3As are supported by various educational organisations and public initiatives. Certainly, one of the most important associations in this regard is the Polish Association of Universities of the Third Age, which was founded in 2007, and which collaborates with other international U3As, in Austria, Lithuania, Ukraine and Belarus, among others (Polish Association of the Universities of the Third Age, n.d.). There are three types of U3As in Poland—namely, those housed within university structures, others which are incorporated in local government authorities, and others serving as self-directed non-governmental organisations.

Legal Structure

The majority of U3As are independent non-governmental organisations. Their functioning is governed by the Non-Governmental Organisations Act, the Foundations Act, and the Public Benefits and Volunteer Work Act. These organisations set their governing board and auditing body according to procedures set out in their respective statuses. In most instances, U3As who serve as NGOs tend to request universities to award them academic patronage or special cooperation agreements (Gołdys et al., 2012). As a result, they very often receive noticeable support from local and regional governments (such as facilities, office equipment, and logistic assistance). Other forms of support are provided through the Public Benefits and Volunteer Work Act, whereby local governments are obliged to develop and coordinate programmes of collaboration with non-governmental organisations. U3As which are part of academic organisations—both public and non-public—are established by university authorities and managed by rectors' representatives. Hence, whilst parent universities determine the rules of functioning, and support the U3As in organising and carrying out their learning activities, it also follows that such U3As are structurally and financially dependent on such traditional universities. U3As that function within the structures of local government authorities (e.g. cultural centres or libraries) are established by resident committees and managed by paid personnel. Local authorities determine how such U3As are legislated and provide both logistic and financial support for the learning activities, with members paying the set tuition fees (Borczyk, Nalepa, Knapik, & Knapik, 2012). In most instances, the location where the U3As are situated determines their legal structure. Whilst in large cities (100,000–1,000,000 residents) most U3As form part of universities, in middle-sized cities (50,000–100,000 residents) they are generally set up as non-governmental organisations. In small towns and villages (up to 20,000 residents), U3As are most often established in local cultural centres. In 2015, the majority of U3As formed part of an association or a foundation (57%), almost one in four (23%) were under the auspices of a traditional university, and 16% are based in cultural centres and libraries (Głowny Urząd Statystyczny, 2016).

U3A Financing

U3As in Poland are mainly self-financed (membership contributions and fees), though many U3As necessitate an entry fee, as research shows that half of the universities charge tuition fees (monthly, per semester, or annually). Financial support is also offered by regional authorities and cultural institutions. Detailed analyses of key U3A funding sources revealed the following percentages of income: 35% from tuition fees, 25% from subsidies from local governments (regional, city, municipal), 18% from membership fees, 11% from institutional sponsorships within which U3A operate, 8% from public-government grants, and 3% from other funds (Głowny Urząd Statystyczny, 2016). At the same time, U3As are only able to make ends meet through voluntary work on behalf of their membership body.

Operating Standards

Legal and organisational regulations governing U3As are highly diverse and not subject to any unified or central standards (Borczyk et al., 2012). In 2012, Polish Association of Universities of the Third Age coordinated the project *Professional Universities of the Third Age in Poland* to develop operational standards for U3As, and implementation guidelines on how these can be promoted and put into practice. The emergent standards covered obligatory requirements of Polish law, as well as additional requirements regarding curricula and administration, and standards of collaboration with third parties. However, their implementation is at the sole discretion of each U3A although it was also premised that the legal functioning of U3As should be based on valid lawful acts and regulations governing functioning of non-governmental associations and foundations. Substantive standards included, among others, educational and activation programmes. Curricula are generally developed by U3A coordinators in liaison with student committees, and academic councils that include representatives of universities, scientists, academic experts, and leaders in local authorities. Each learning programme generally consists of the title, lecturer's profile, place, date, and time of classes (Hrapkiewicz, 2009). A standard catalogue of learning programmes is universally developed (Borczyk et al., 2012). At the time of writing, older adult education is directly related to different aspects of their life such as physical and mental health, social and financial security, social integration and participation, increased participation in life of community, family and friends, and personal development: optimism, motivation, leisure, effective use of time (Escuder Mollon, 2014). There are also standards of cooperation of U3As with different institutions: public (e.g. state and local government units, universities and schools), private, media, and other NGOs. This cooperation is established to carry out different joint activities (Borczyk et al., 2012).

Organisation

U3As are usually large institutions with several hundred members, but there are also smaller units with several dozen students. The diversity of teaching and learning styles is generally twofold. The first is highly academic and almost formal in nature. The academic year is divided into years and terms (semesters) and involves participation in various forms of classes, where members are required to notify their absence in advance to receive a certificate of participation (completion of a term depends on participation in obligatory classes, and a minimum entry requirement of either a baccalaureate or high school diploma). A second teaching and learning style constitute of open formula meetings, usually adopted by U3As in cultural centres, welfare institutions, and small-town municipalities. This variety means that U3As in Poland cannot be pigeonholed into a single *modus operandi*, and are characterised by constant changes in their range and forms of activities, typified by diverse programmes and geragogical practices (Szarota, 2008). However, some U3As run 'lighter' classes and numerous events that provide their members with opportunities to further develop their social relations, sometimes in conjunction with a social and voluntary project. As one U3A member attested,

> … we have a lot, really a lot [of activities]…we have these lectures, which really embrace various topics ranging from politics to health…all themes are somehow incorporated… moreover, there are many hobby groups that if one…a week is too short in order to …I attended English, computer science…I was going to the swimming pool…I took part in the excursions, which were organized…well, as I said, it is not possible to attend everything.
>
> U3A member, cited in Wilińiska, 2012: 295

The academic year is divided into two terms, from October to mid-June, with classes usually held two or three times weekly, and consisting of lectures, seminars; language lab forums; hands-on computer sessions; workshops; recreational activities held in gymnastic halls, swimming pools or country walking trails; and hiking trips, bus tours, and sport games (Borczyk et al., 2012; Gołdys et al., 2012; Tomczyk, 2012).

Member Profile

Research conducted among U3A members identified two types of members. Whilst 'comfort-seekers' take interest in this form of education because they can benefit from attractive and often cheaper offer of activities and courses, 'active' members contribute to what is going on at the U3A (Gołdys et al., 2012; Kotler, 2002). Although U3As in Poland are often marketed to retired persons (namely, persons above statutory retirement age), organisers do not impose an age requirement for membership (Szarota, 2008). The range of members in U3As in Poland is from as little as 10 participants to over 2,000. As elsewhere, 85% of participants are women, with the majority being in the 60–65 age cohort. However, among members aged 50 and less, the number of men exceeded that of women as it totalled 63% (Głowny Urząd Statystyczny, 2016). The causes underlying this majority of older women in the U3A membership body include the following: longer life span of women, better health and general well-being levels among women, women being more inclined to participate in social activities whilst men prefer more individual forms of activity, a larger pool of older women as many never joined the formal labour market, a personalised attitude towards retirement ('time for myself'), and the fact that classes offered by U3As tend to address women's, rather than, men's needs and interest (ibid.). Most of the male members tend to be the husbands of female participants, and tend to participate in U3A activities during excursions where they accompany their wives (Gołdys et al., 2012).

Most U3A members tend to have recently retired or opted for early retirement. The majority are aged between 60 and 70 (62%), 25% are over 70 years old, and 11% are over 75 years old. At the other side of the age continuum, 12% were aged 59 or younger, with 1% aged less than 51 (Głowny Urząd Statystyczny, 2016). On average, 30% of the U3A member population tend to remain enrolled until they are physical unable to continue attending class (Czerniawska, 2013). A significant percentage of U3A members have higher (39%) or upper secondary (49%) education (ibid.) and are often retired teachers, health professionals, or representatives of other professions. In general, U3A members are people who lived their whole life dynamically, and actively engaged in various interests and pastimes (Gołdys et al., 2012). Their main motivation to enrol in U3As is to continue their personal socio-psychological and intellectual development (Halicki, 2014). As expected, this gender and socio-economic status imbalance determines the type and range of subjects covered during classes, so that the curricula at U3As is highly biased in favour of those interests generally held by ageing women and middle-class elders. This situation strengthens the process of social exclusion of men and older persons from working-class backgrounds: 'the bourgeois ideals implicit in the activities and rhetoric of Universities of the Third Age, as well as the bodily health required to attend, can exclude many older people from access to a culturally-sanctioned good old age' (Robbins-Ruszlowski, 2013: 165). Hence, when reflecting about the state of third age education in Poland, it follows that U3As do not represent the whole spectrum of Polish seniors

(Lipka-Szostak, 2013). Indeed, research indicated that over 90% of U3A members identified with activities that included intellectual games, willingness to learn, curiosity regarding the world and other people, and physical activity (Fabis, Muszynski, Tomczyk, & Zralek, 2014). Other characteristics—mentioned by over 83% of respondents—included an interest in culture and art, independent thinking and opinions, and ability to deal with difficult situations (ibid.). Among personal traits, more than 65% of students listed honesty, independence, reliability, and diligence. Other characteristics identified by over half of respondents include being organised, sensitive, optimistic, realist, social, resourceful, balanced, and intelligent (ibid.).

U3A and Quality of Life

U3As in Poland contribute significantly to improving the quality of life of the country's older cohort. As learning institutions, albeit of a non-formal type, they act as catalysts towards better levels of physical, social, and psychological well-being for its members. U3As bid older persons in Poland the opportunity to use their free time, offering different forms of intellectual engagement that mitigates against the monotony of retired life. In a study which specifically asked Polish U3A members how their engagement in learning impacted their quality of life, Maćkowicz & Wnęk-Gozdek concluded as follows:

> …the participation of seniors in U3A classes is beneficial for the quality of their lives. U3A classes prevent isolation and result in positive changes of both an individual and social dimension and generates long-term benefits… Participation in various forms of educational activity broadens one's horizons. Seniors look to broaden their knowledge and circle of interests, drawing from a huge pool of opportunities. They often discover their passions and potential anew. The benefits brought by their activity are priceless, especially in the context of the ageing of society…
>
> Maćkowicz & Wnęk-Gozdek, 2016: 196

Other research demonstrated that U3A members, both in times of their youth and at present, are more intellectually and physically active than non-members (Bicka & Kozdroń, 2003; Grzanka-Tykwińska, Chudzińska, Podhorecka, & Kędziora-Kornatowska, 2015; Maćkowicz & Wnęk-Gozdek, 2015). Indeed, U3A members are very physically active, with as much as 87% of U3A members performing some kind of gymnastic exercises outside learning hours (Ziębińska, 2010). U3As also engage their members in social relations, offering much potential in improving their 'bridging' and 'bonding' levels of social capital—which also functions in the prevention of exclusion and marginalisation in later life (Maćkowicz & Wnęk-Gozdek, 2016). Indeed, in a study on Polish U3As framed in the context of national historical politics and public conversations about old age, Robbins-Ruszlowski found that

> …women who attend Universities of the Third Age can be understood as challenging dominant stereotypes of old age in general, and of old women in particular…Institutional focus

on Poland's membership in the EU can be understood as an attempt to overcome negative associations of older people with the socialist past, while learning new hobbies and skills can be understood as an attempt to overcome negative stereotypes of old age as a time of decay.

Robbins-Ruszlowski, 2013: 165

U3As also allow members to assume new social roles through the voluntary work available at the centres, with research confirming that at least 10% of members engaged in some kind of voluntary activity (Ziębinska, 2010). U3As also enable older persons to gain new competencies necessary for daily life in the contemporary information society and overcome their fear of new technologies. Programmes include courses in information and communication technology, word processing, and the Internet, even covering areas as e-mailing, online shopping, online pharmacies, payment online, mobile phoning, and online banking (Szarota, 2008). The feedback is encouraging as 70% of U3A members reported that the computer and Internet skills they learnt are extremely helpful in their daily life (Ziębińska, 2010). Whilst there is no doubt that U3As enhance the quality of life of their members, one must not overlook that many Polish older persons, as in other countries (Formosa, 2012), who are neither women nor middle-class, meet various obstacles in their attempt to enrol themselves and participate in U3As:

> Despite the undeniably positive influence on the quality of life among seniors the U3A is still not accessible for the average senior because of their location, their material conditions, and their individual readiness to take part in such activities. In Poland there is also a very strong tradition of engaging seniors in the care of their grandchildren, and this in many cases makes it impossible for them to take part in any self-development or self-actualization programs.
>
> Maćkowicz & Wnęk-Gozdek, 2016: 186

Hence, it is augured that the U3A movement in Poland should spend some of their energies in reaching older persons who are not typical members, especially those living in small towns and villages, who tend to experience social isolation and marginalisation.

Challenges for the Future

The later stages of the life course are fraught with many challenges, as older persons do not only require learning opportunities to keep their mind active, but also require support in maintaining a strong social network, and personal assistance in times of duress and illness. Hence, it is important that U3A associations widen their focus from strictly learning enterprises to embrace a more welfare stance that keeps touch with their members as they experience transitions and live events related to old age. Moreover, it is high time that U3As come out from their comfort zone in organising learning programmes in the arts and humanities and include new subject areas such as financial literacy, economics and marketing, active citizenship, care services for frail elders and/or elders with disability, human rights of older people, and counselling (Ministerstwo Edukacji Narodowej, 2013). It is also perhaps time to include subjects which, for many years were considered taboo, such as palliative care, death and dying, and how to mitigate the problems of widowhood (Gołdys et al., 2012). At present, U3A members tend to be surprised and incredulous when they encounter disease or death. In the words of one Polish U3A member,

> It happened to us on New Year's Eve. We were invited by a friend, four of us. We went and no one opened the main door for us…there were four families living on the same floor so we rang the bell to a neighbour. She opened the door for us; after knocking on our friend's door, we heard that she could not stand up; once we called the ambulance, we learned that she had a stroke; she has been bed-ridden since then, what will happen next? No one knows…we call and visit this sick one.
>
> U3A member, as cited in Wilińska, 2012: 295

Interest in information and communication technology is also becoming increasingly popular among Polish older persons, since with each passing year an increasing number is gaining more and more access to online forms of communication (Kędziora-Komatowska & Grzanka-Tykwińska, 2011). In the immediate future, older adult learning will be progressively hinged on e-learning methodologies (Formosa, 2009), especially when the target group consists of older persons living in remote rural areas or who experience mobility difficulties (Grzanka-Tykwińska et al., 2015). As already hinted at in a previous section, U3As are warranted to develop their academic and social programmes in a way that attract more elder men, and more older persons with disability, even if this requires the organisation of transport and the engagement of volunteers for assistance (Gołdys et al., 2012). Another challenge includes the availability of continuous professional development programmes for the 'educators' who are responsible for the planning and facilitating of lectures at the U3As. As various literature attests, teaching older adults and learning styles in later life require specific geragogical skills on behalf of 'educators' that need to be cultured in an academic environment by expert personnel trained in educational gerontology. Finally, there is also a need for U3As to engage in intergenerational programmes by planning learning projects with younger adults, youth, and even children (Borczyk, 2012).

As more countries experience the coming of population ageing, U3As can take on a more active role in supporting their cognitive and physical independence, as well as transforming the contemporary environment in ways that are advantageous to older persons (Battersby, 1985; Formosa, 2014, 2016). Late-life learning has the potential to mitigate against learned helplessness in later life and to provide older persons experiencing retirement and an empty nest with a renewed sense of purpose. As Majewska-Kafarowska (2009) argued, in many ways U3As constitute a vehicle for the further development of one's 'becoming', an opportunity for growth and resilience in a new stage of the life course.

References

Battersby, D. (1985). Education in later life. What does it mean? *Convergence, 18*(1–2), 75–81.
Bielowska, K. (2015). Profesor Halina Szwarc (5.5.1923–28.5.2002). *Biuletyn, 15*, 5–8. https://www.ur.edu.pl/file/90261/srodki_UTW_15_poglad+%282%29.pdf. Accessed March 12, 2018.
Bicka, A., & Kozdroń, E. (2003). Aktywność ruchowa w stylu życia ludzi starszych. *Annals Universitatis Mariae Curie-Sklodowska, 58*, 61–65.
Borczyk, W. (2012). *Sytuacja osób starszych w kontekście doświadczeń Uniwersytetów Trzeciego Wieku*. http://www.wrzos.org.pl/download/Ekspertyza_2_ASOS.pdf. Accessed February 22, 2018.
Borczyk, W., Nalepa, W., Knapik, B., & Knapik, W. (2012). *Standardy działania uniwersytetów trzeciego wieku*. Nowy Sącz: Wyd. Ogólnopolska Federacja Stowarzyszeń UTW.
Czerniawska, O. (2013). Uniwersytet Trzeciego Wieku 30 lat działania. Przemiany, dylematy i oczekiwania w epoce ponowoczesnej. *Chowanna, 2*(33), 97–114.
Escuder Mollon, P. (2014). Quality of life. In P. Escuder Mollon & A. Gil (Eds.), *Edukacja a jakosc zycia seniorów* (pp. 13–28). Czestochowa: Wydawnictwo Akademii im. Jana Długosza w Czestochowie. http://www.edusenior.eu/data/outcomes/wp5/EduSenior-guide-PL.pdf. Accessed March 12, 2018.
Fabis, A., Muszynski, M., Tomczyk, L., & Zralek, M. (2014). *Starosc w Polsce. Aspekty społeczne i edukacyjne*. Oswiecim: PWSZ w Oświecimiu.
Formosa, M. (2009). Renewing universities of the third age: Challenges and visions for the future. *Recerca, 9*, 171–196.
Formosa, M. (2012). Education and older adults at the University of the Third Age. *Educational Gerontology, 38*(2), 114–126.
Formosa, M. (2014). Four decades of Universities of the Third Age: Past, present, and future. *Ageing & Society, 34*(1), 42–66.
Formosa, M. (2016). Malta. In B. Findsen & M. Formosa (Eds.), *International perspectives on older adult education: Research, policies, practices* (pp. 161–272). Cham, Switzerland: Springer.
Głowny Urząd Statystyczny. (2016). *Uniwersytety trzeciego wieku—wstępne wyniki badania za rok 2014/2015*. https://stat.gov.pl/files/gfx/portalinformacyjny/pl/defaultaktualnosci/5488/10/1/1/uniwersytety_trzeciego_wieku.pdf. Accessed January 19, 2018.
Gołdys, A. et al. (2012). *ZUM na UTW. Raport z badania*. Warszawa: Towarzystwo Inicjatyw Twórczych "ę".
Grzanka-Tykwińska, A., Chudzińska, M., Podhorecka, M., & Kędziora-Kornatowska, K. (2015). Uniwersytety Trzeciego Wieku wczoraj, dziś i jutro. *Gerontologia Polska, 4*, 165–168.
Halicki, J. (2014). Zaspokajanie potrzeb edukacyjnych jako czynnik aktywnego starzenia się. In P. Szukalski, B Szatur-Jaworska (Eds.), *Aktywne starzenie się. Przeciwdziałanie barierom* (pp. 142–151). Łódź: Wydawnictwo Uniwersytetu Łódzkiego.
Hrapkiewicz, H. (2009). Uniwersytety Trzeciego Wieku jako jedna z form kształcenia osób starszych. *Chowanna, 33*, 115–127.
Kędziora-Kornatowska, K., & Grzanka-Tykwińska, A. (2011). Osoby starsze w społeczeństwie informacyjnym. *Gerontologia Polska, 19*, 107–112.
Klimczuk, A. (2013). Kierunki rozwoju uniwersytetów trzeciego wieku w Polsce. *E-mentor, 4*(51), 72–77.
Kotler, P. (2002). *Marketing. Podręcznik Europejski*. Warszawa: Polskie Wydawnictwo Ekonomiczne.
Lipka-Szostak, K. (2013). Seniorzy—trendy, wyzwania, działania. In K. Lipka-Szostak (ed.) *Edukacja osób starszych. Uwarunkowania. Trendy. Metody* (pp. 47–60). Warszawa: Stowarzyszenie Trenerów Organizacji Pozarządowych. https://ec.europa.eu/epale/sites/epale/files/edukacja_osob_starszych_-_publikacja.pdf. Accessed March 12, 2018.
Maćkowicz, J., & Wnek-Gozdek, J. (2015). 'It's never too late to learn'—How does the Polish U3A change the quality of life for seniors? *Educational Gerontology, 42*(3), 186–197.

Maćkowicz, J., & Wnek-Gozdek, J. (2016). Transformacja starszych kobiet pod wpływem odd-ziaływań edukacyjnych uniwersytetu trzeciego wieku—Studium przypadku. *E-mentor, 2*(64), 45–55.

Majewska-Kafarowska, A. (2009). Edukacja (seniorów?) w procesie rewalidacji społecznej starzejącego się społeczeństwa. *Chowanna, 2*(33), 213–224.

Ministerstwo Edukacji Narodowej. (2013). *Perspektywa uczenia się przez cale życie.* www.men. gov.pl/index.php/uczenie-sie-przez-calezycie/770-perspektywa-uczenia-sie-przez-cale-zycie. Accessed February 25, 2018.

Ministerstwo pracy i Polityki Społecznej. (2014). *Założenia Długofalowej Polityki Senioralnej w Polsce na lata 2014–2020.* https://www.mpips.gov.pl/download/gfx/mpips/pl/defaultopisy/8489/ 1/1/ZDPS%2014-02-04%20%20Monitor%20Polski.pdf. Accessed April 14, 2018.

Polish Association of the Universities of the Third Age. (n.d.). *Polish Association of the Universities of the Third Age.* https://www.federacjautw.pl/. Accessed June 19, 2017.

Polskie Towarzystwo Gerontologiczne. (2017). *Halina Szwarc.* http://gerontologia.org.pl/wp-content/uploads/2017/06/Prof.-dr-med.-Halina-Szwarc.pdf. Accessed February 25, 2018.

Półturzycki, J. (2013). W 90. rocznicę urodzin—Profesor Halina Szwarc—twórczyni uniwersytetów trzeciego wieku w Polsce. *Rocznik Andragogiczny, 20,* 265–277.

Robbins-Ruszkowski, J.-C. (2013). Challenging marginalization at the Universities of the Third Age in Poland. *Anthropology & Aging Quarterly, 34*(2), 157–169.

Szarota, Z. (2008). Przestrzeń edukacyjna Uniwersytetu Trzeciego Wieku, *E-mentor, 3*(25). http:// www.e-mentor.edu.pl/artykul/index/numer/25/id/559. Accessed January 11, 2018.

Szarota, Z. (2009). Seniorzy w przestrzeni kulturalno-edukacyjnej społeczeństwa wiedzy. *Chowanna, 2*(33), 77–96.

Szarota, Z. (2015). Senior policy in Poland: Compensation of needs and active ageing. *Polish Social Gerontology Journal, 2*(10), 99–112.

Tomczyk, Ł. (2012). Education of older people in the field of information technology on the example of Polish Universities of the Third Age. *Procedia—Social and Behavioral Sciences, 55,* 485–491.

Wilińska, M. (2012). Is there a place for an ageing subject? Stories of ageing at the University of the Third Age in Poland. *Sociology, 46*(2), 290–305.

Ziębińska, B. (2010). *Uniwersytet Trzeciego Wieku jako instytucje przeciwdziałające marginalizacji osób starszych.* Katowice: Uniwersytet Śląski.

Jolanta Maćkowicz holds the post of Associate Professor at the Pedagogical University of Cracow in Poland and a Visiting Scholar at various European universities. She is an accomplished researcher in the educational, social and gerontological fields and is the author of about 50 publications (indexed in 'Journal Citation Reports') as well as reviewer of articles submitted in journals (indexed in 'Journal Citation Reports'). She is the national representative of the International Network for the Prevention of Elder Abuse (INPEA) and independent expert reviewer and rapporteur of European Commission Research Programme HORIZON 2020.

Joanna Wnęk-Gozdek is Assistant Professor at the Pedagogical University of Cracow in Poland within the Institute of Educational Sciences. Her academic interests are focused on social gerontology and pedagogy, as well as contemporary didactics related to information and communication technologies. Her is the author of over 60 publications (indexed in 'Journal Citation Reports'), and member of the Social Pedagogy Team at the Pedagogical Sciences Committee of the Polish Academy of Sciences and Polish Association of Social Gerontologists.

Chapter 9
Universities of the Third Age: Learning Opportunities in the Russian Federation

Gulnara Minnigaleeva

Introduction

The number of older persons is growing rapidly the world over, and Russia (Russian Federation) is no exception. According to projections by the Russian statistics office, the percentage of persons aged 60-plus will reach over 30% of the country's population by the year 2030 (Russian Statistics Office, 2018). Here, it is noteworthy to point out that in 2018, the mandatory retirement ages for women and men in Russia were 55 and 60, respectively. Ageing policies are discussed and updated both at federal and national levels, and whilst it is hoped that the supply of ageing services meets demand, it is positive to note that the number of services and opportunities available for older adults is increasing in Russia. Lifelong learning opportunities comprise a notable segment of available services. Education for older adults is organised in different forms, ranging from informal to formal. An increasing number of institutions are becoming interested in providing learning opportunities for older adults, and host associations include city authorities, universities, governmental social service centres, non-profit organisations, and individuals acting in their personal capacity. So far, the role of Universities of the Third Age (U3A) in Russia has been more that of engaging seniors in active lifestyles, assisting them to socialise with same aged peers, and maintaining their mental health, rather than educating them for professional qualifications. However, following the recent pension reform which will increase the pensionable age from 55 to 60 for women and from 60 to 65 for men by 2028, it is envisaged that the situation may change towards an increasing presence of learning options targeting professional development and new careers in older age.

G. Minnigaleeva (✉)
National State Research University Higher School of Economics, Moscow, Russia
e-mail: gminnigaleeva@gmail.com

© Springer Nature Switzerland AG 2019
M. Formosa (ed.), *The University of the Third Age and Active Ageing*, International Perspectives on Aging 23, https://doi.org/10.1007/978-3-030-21515-6_9

Older Students and Motivation

Older people in Russia are generally willing to stay active in retirement. Research conducted in 2014 reported that older adults were most willing to engage in different activities between the ages of 63 and 71 (Levintov, 2015). Whilst 81% of respondents falling within this age cohort stated that they wished to remain engaged in some kind of entertaining physical or mental activity, the most popular activities were visiting their dacha (32%), handcrafts (22%), reading (20%), and hunting and fishing activities (11%). Willingness to remain engaged in older adult learning decreased with age, with the largest percentage of people interested in learning being at pre-retirement and early retirement stages—37% between the ages of 58 and 62, followed by 25% between the ages of 63 and 71, and 11% at the age of 72 and older. Another research study, however, found that only 14–20% of persons aged 50 and over would like to participate in a learning programme to learn new skills or knowledge (Barsukov, 2017). It could be that results depend on the region, particular economic and cultural situations, and the affordability of learning programmes. Another research study demonstrated that only 17% of older persons within the 55–64-year-old cohort participated in any form of lifelong learning, and only 4% engaged in some kind of formal education (Bondarenko, 2017). Such differences could be the result of insufficient levels of learning opportunities or difficulties of accessing existing programmes. As expected, more persons aged 55-plus than in any other age group choose to study because of a personal interest, as a hobby, or for leisure (ibid.).

A survey of current participants of one U3A reported that whilst 85% enrolled to learn new things, some 40% were attending to seek the company of peers, and up to 80% were searching for a change in lifestyle (Sidorchuk, 2013). Most older adults who enrol in U3As are women; this is not surprising given that life expectancy in Russia is only 66 years for men, and as high as 77 years for women. Reflecting other international studies (Formosa, 2014), about 80% of members of the Oryol U3A are women (Valerianova, 2014); and at the U3A of Mariy State University women comprise as much as 91% of all members, with 66 being the average age of all members (Domracheva, Morova, Makarova, & Mal'ceva, 2016). Moreover, members tend to have higher-than-average levels of educational attainment and qualifications, as 62% and 38% of members at the latter U3A possessed higher education qualifications and professional college degrees, respectively (ibid.). Whilst list of motivations is not exhaustive, the most popular reasons included the need for knowledge, positive emotions, personal growth, and typical personality of facilitators (Minnigaleeva, 2013). Indeed, it was clear that most participation in older adult learning had little to do with employability, and most learners (72%) did not expect to receive any kind of certification following the completion of their studies (ibid.). Most older learners were not seeking education opportunities for employment purposes, but their key motivations included a desire to improve their quality of life and social well-being.

Policies

In 2016, the Russian government adopted the *Federal Strategy for Older Persons* (Government of Russia, 2016) that promoted and safeguarded the interests of older citizens. The Strategy defined goals, principles, tasks, and priorities of the governmental social policies towards older citizens for the period 2012–2018. Besides others, the key goals of the Strategy included providing access for older citizens to informational and educational resources, promoting employment opportunities for older persons, and raising the level of financial competence in later life. The Strategy was created through cooperation between the government and representatives of civil society groups, educators, and non-governmental organisations, and therefore highlights the needs and priorities of older persons in Russia. It referred to many different types of educational activities and focused on creating conditions for raising the levels of financial and legal literacy; development of cultural centres; increase of volunteering roles in older adult learning; improving access to information for older people; and the development of accredited learning plans for older adults.

The adoption of the Strategy was a highly positive step towards the improvement of active and successful ageing in Russia, especially when one considers that the preceding decade witnessed no government-set priorities in the interest of older persons. The previous Federal special purpose programme consisted of the *Older Generation* programme which was launched in 1997 and terminated in 2004. The *Older Generation* programme targeted the provision of basic needs of older persons following the harsh economic climate that characterised Russia in the 1990s. Unfortunately, learning was neither considered nor mentioned in the objectives of this programme. Although there was no national policy towards older adults until 2016, it is positive to note that local and regional authorities still initiated welfare programmes for older people, including educational ones. For example, a region-wide governmentally sponsored programme—'Third Age Universities for all'—came into operation in 2011 in the Republic of Bashkortostan, providing complimentary learning opportunities to thousands of older adults. Whilst the programme had significant limitations, it set an important precedent and has been replicated in many other regions. Such initiatives have attracted hundreds of older students and, when funding and resources permitted, both the number of participants and range of courses increased in a short matter of time (Authorities of the Jewish Autonomous Region, 2018).

History

For many years, lifelong learning was a part of the official education system in the Soviet Union. *Znaniye* (meaning knowledge) is an association that was founded in 1947; it became one of the Soviet Union's pillars of educational policy. During the Soviet times, *Znaniye* was highly supported by the government, had branches all over the Soviet Union, and successfully disseminated progressive scientific knowledge from experts to ordinary people. Although for many years, the work of *Znaniye* towards older adult learning was limited, this changed following the political and economic changes of the 1990s, as several departments of *Znaniye* initiated various study programmes for older people. One of the pioneers, Tatyana Kononygina, launched a U3A in the city of Oryol in 1997 (Oblast Branch of Association *Znaniye* of Russia, n.d.). As older people often felt lost in the changing socio-economic conditions of the time, and this U3A offered them a new lease of life, it quickly gained much popularity. The first announcement in the local newspaper attracted 700 older students. Their interests mostly revolved around computer literacy; gardening (as most were surviving by growing their own vegetables); and most importantly, socialising (Tolkacheva, 2018). Following this successful project, more U3As were created by other *Znaniye* branches. In 2003, the School of the Third Age was founded in St Petersburg. This school was highly reminiscent of the Retirement Institutes and the Folk (People's) Schools in North America and Northern Europe, respectively. The St. Petersburg School's special feature was the distinctive role of older persons in forming its curriculum and social events (Levintov, 2015).

After the millennium, governmental social service centres also began offering learning opportunities for older persons. Such centres held free venues provided by the government and invited speakers to give public talks at these premises. Since at the early 2000s, Russia was still characterised by an unstable economy, most lectures were based on financial literacy, sales of all kinds of food supplements, and 'miraculous' health boosters. Nevertheless, such practice functioned to legitimise the practice of providing learning programmes to older adults, and in the long run, had an immense positive contribution on future developments in older adult learning in Russia (Minnigaleeva, 2004). In the second half of the 2000s, more non-governmental organisations and volunteer associations were establishing learning groups for older persons in different regions of the country. In some regions, local and regional authorities even supported such initiatives by allocating a special financial budget for third age learning.

Current Third Age Learning Prospects in Russia

Nowadays, older adult learning in Russia is a 'hot' topic. It is supported by government policies and has become quite visible in the press and popular life. There is no single organiser or just one leading association in the field. The range of organisers of late-life learning possibilities include universities, governmental social service centres, individual initiatives, local and regional authorities, libraries, and non-profit organisations and associations. Whilst some of them do bear the U3A tag, others have titles like 'Pensioners' university', 'Senior school', 'People's university for pensioners', 'Silver university', 'Golden University', 'Hobby groups for elders', 'Learning clubs', and so on. These programmes are enjoyed by older learners and are believed to be major contributors to active ageing in Russia. Most of them are informal settings, with an absence of any official diploma or certificates, where social inclusion and the well-being of older persons are heralded as the key objectives.

In recent decades, there have been more opportunities for older adult learning. Governmental social service centres were founded in various cities in Russia between the mid-1990s and mid-2000s. As their target audience included vulnerable people—especially people with disability and older people—most of their members were aged over 60. The primary purpose of the centres was to provide for the basic needs of the target audience; for example, offering meals, providing clothes, and basic medical services like blood pressure checks. Many of them also tried to cater to other needs of older adults such as socialising, organising leisure activities, and of direct interest to this chapter, learning activities. Thus, hobby groups, lectures on different topics and craft classes were organised either by the centres' employees or other individuals who were willing to volunteer as facilitators. The centres often employed psychologists and cultural events organisers and were also ready to provide space for internal and external volunteers. The primary function of social service centres, still in place today, is the support of older people's basic needs and the servicing of low income and low mobility persons and vulnerable groups in general. As a result, these centres suffer stigma, which sometimes causes difficulties with attracting older adults to attend the programmes. Moreover, non-governmental organisations also play a very important role in providing learning opportunities for older people. *Znaniye*, which pioneered the field of older adult education in Russia, continued growing and developing its services to different regions of the country throughout the 2000s, although no branch follows the same strategies, focus, and range of activities. For example, the Krasnoyarsk branch relied on practices forwarded by the German U3A association, which coordinated the European Union's project on 'Broadening of opportunities of older persons' participation in social and political process in democratic development of Russia'. The classes focused on the community participation of older persons, creativity, broadening use of information technologies, biography methods, and practical advice on living alone (Ovchinnikov, n.d.). As reported on Znaniye's (2018) website, U3As are currently run in 16 different cities.

The Union of Pensioners, which is another Russian non-governmental organisation, was one of the first societies which started a formal collaboration with a university in Kazan in 2007. Computer literacy courses were most popular from the very beginning, and today the Union holds annual country-wide competitions on computer knowledge among retirees. These competitions are quite popular, attracting thousands of participants and encouraging older persons to study computers even more intensely. Grass-root organisations also organise multi-purpose learning programmes. In such organisations, there generally is a range of other programmes for older persons ranging from volunteering and hobby groups to advocacy groups. Moreover, their learning programmes are more diverse than those in government-sponsored programmes. They may include somewhat unusual learning programmes such as the organisation of television show productions, journalism, and poetry. At the same time, many U3As are being set up by Veterans' councils. In most regions, Veterans' councils have an advantage of free use of premises granted by local governments. Some educational opportunities for older adults emerge as a result of individual initiatives. Several learning programmes in information and communication technology (ICT) for older adults can be found on the Russian Internet (Computer academy for pensioners, n.d.). Clearly, at some point in time they served as a good start for older learners trying to navigate the Internet for the first time in their lives. It is not uncommon that individuals organise face-to-face learning groups and may be also found cooperating with governmental social service centres, Veterans' councils, or libraries to host their activities.

City authorities have also been investing substantially in lifelong education. For example, third age learning is a part of the special programme *Moscow Long life*, which was launched by the local Moscow government in 2018, and the *Silver University* project which was launched by the Moscow Government University and which to-date has already served 650 older persons in over 54 groups. Several study programmes run at the latter project constitute certified professional programmes, and all Moscovites of pensionable age are eligible for membership (Moscow Silver University, n.d.). Social service centres accept applications for the university, although some study programmes such as 'English for guides' and 'Smartphones for everyday life' require preliminary assessment (Moscow Government, 2018). A new senior career centre sponsored by the city government has just opened its doors in Moscow; its aim is to provide vocational-oriented education which would allow alumni to find suitable work to earn additional income to their pensions. The most recent course targeted real estate agents (Silver Age, n.d.). As mentioned earlier, universities and other extensive formal institutions participate in creating learning opportunities for older populations, but rarely initiate them on their own. For example, in 2015 the National Research University of Informational Technologies, Mechanics and Optics (n.d.) introduced several courses on their online study programme. In this case, the non-profit partner was the charity organisation 'Nevsky Angel', which helped to engage and educate volunteers (Nevsky Angel, n.d.). Collaborations are common among U3As in Russia as most educational projects for older people are organised in partnerships. Governmental organisations often invite universities, volunteers, or non-profit organisations to set up study programmes or public lectures on their premises. Alternatively, non-profit organisations request using space in universities, or other educational institutions, such as secondary schools or art schools, and even ask teachers to volunteer for their U3A projects. Many non-profit organisations urge older persons to initiate their own courses with the effect that in some cases secondary school or university teachers volunteer to teach older adults, or that retirees with no previous teaching experience start sharing their expertise or skills in learning circles aka hobby groups (Wisdom Ripening, n.d.).

Concepts and Curricula

What is common for almost all Russian U3As is that the type of learning is informal, with no graduation or degree conferred, no exams, and no diplomas. Some U3As corroborate with regular universities or other educational institutions to issue certificates for the coursework, but that is the extent of it. The primary purpose is to keep people active and engaged in social life, as well as to help them develop personal skills. Many learners also report that these programmes help them fulfil dreams they could not fulfil whilst they were fully committed to raising their children and focusing on job and career development. Forms of tutoring vary depending on the host organisation. Whilst universities provide more formal ways of education—through lectures and seminars—social service centres organise public talks, discussions, and hobby groups. Overall, there is an understanding that older persons prefer a more active type of learning than a passive receipt of information. Researchers also note that older students enjoy making presentations and role-plays rather than repetitive exercises (Afanas'eva & Savina, 2015). More often than not, it is non-profit organisations which prefer more active engagement of older persons through volunteering and the creation of their own educational curricula (Wisdom Ripening, n.d.; House of Projects, n.d.). Based on a U3A project of Shuya regional government financed by Ladoga Foundation, Belov (2014) formulated six principles for the teaching of older adults—namely, the (i) principle of joint individual and group study process; (ii) principle of systematic learning; (iii) principle of full understanding of the learning process, (iv) principle of using the life experience for learning; (v) principle of individual election of learning strategies; and (vi) principle of socialising whilst learning.

Considering that the vast variety of institutions involved in U3A learning and the fact that there is no singular agenda for all of them, it is expected that the curricula vary greatly. Nevertheless, it is possible to identify some consistent patterns. Analyses of study programmes and hobby groups for older learners in the Krasnoyarsk region (Ovchinnikov, n.d.) demonstrated that courses can be divided into five distinct groups namely: social and psychological targeting mental health and relationships; medical and healthy life style with public lectures on changes in human body and easy remedies; legal concerning questions of older persons' social rights and other legal issues such as succession rights; culture and history which, besides regular classes, may include theatre going and excursions; and gardening. Such a range of courses is typical for U3As in Russia, together with religious studies, arts, and music. As commonly known, religion in the former Soviet Union was banned from social life, and today many people of different ages aspire to return to it. Spirituality becomes especially important in older age, so many institutions offering learning opportunities for older adults also provide classes in religious studies or philosophy (Union of Pensioners of Russia for the Republic of Tatarstan, n.d. [a] [b]). Languages—especially English—are also common in the curricula, with many learners opting for this language to be able to communicate with their children abroad, or because they want to keep up with the modern world and its common use of English in every sphere. Study programmes which have never found their way into U3A curricula include natural scientific subjects such as advanced mathematics or nuclear physics. This is probably because most older persons are interested in courses which are life-oriented and which may allow for improvement in their quality of life. U3A members prefer knowledge which can be immediately applied, such as classes in physical training or different kinds of yoga-like classes, or handicrafts, singing and dancing.

The range and type of course curricula also depend on the economy and on the development of society. For instance, in the 1990s and early 2000s when the economic situation was difficult and many people were surviving by growing their own vegetables, lectures on the secrets of successful gardening were most popular. After the thriving of computer use in Russia, which happened around the years 2005–2010, computer literacy courses became the most popular. Today, with the ubiquitous spread of smartphones, U3As are offering courses on how to use mobile applications. In Moscow, 500 people have completed this course over a four-month period (Moscow Government, 2018). Other digital technologies which are characterising our lives also induce U3As to provide courses on online banking and electronic government. Another recent novelty at the Moscow U3A is the 'English for Moscow guide' study programme, which was probably introduced because of the Football World Cup which took place in Moscow in the summer of 2018. An interesting study programme which might be of vital importance and highly popular, particularly in Russia, is 'Condominium management'. The job of managing the associations formed by condominium residents usually requires a person to be energetic, highly responsible, available at home, whilst requiring a lot of effort, as it is a volunteer post or at the most a low-paid one. Older persons very often accept this role without the knowledge of the legal side or managing skills. So some regional governments offer these courses to help older persons deal with the emergent challenges. The length of courses also varies to a great extent. The 'Smartphone' study programme in Moscow lasts for 36 h, whilst the professional 'Ticket cashier' study programme totals 160 h. In other cases, language training courses, or handicraft classes, may run for several years and be open for anyone to attend for as long as they would like.

Challenges and Impacts

Despite the fact that U3As have become highly popular and relatively well known among older adults, governments still do not dedicate enough attention to them (Formosa, 2014, 2016). In the Russian context, U3As do not have a steady source of income, and as a result, face problems and complications, even in cases where U3As receive government funding—such as with the 'Third Age Universities for All' project in the Republic of Bashkortostan. In the year 2011, the sum of 2,200,000 Russian Roubles—equivalent to 55,000 Euros at that time—was allocated to the Republic of Bashkortostan for their regional U3A programme. The funding was available in all municipalities of the Republic, divided according to the proportion of older population in the area, and expenses split between the government and institutions. However, such funding only covered the salaries of teachers, and the remaining expenses (for instance, rent of premises and other related expenses) had to be paid for by the relevant institutions, which definitely limited the range of educators able to offer their services. The difficulty with evaluating the effectiveness of U3As also complicates funding decisions as U3As do not have clearly identified goals, in that they are not training seniors for particular professions, and are only involved in non-formal learning. Even for institutions with over a decade of experience in the field of third age universities, it may cause interruption of funding as potential funders do not understand exactly what it is that they are agreeing to finance (Dmitrieva, 2018). Another drawback of many U3As is that facilitators are not prepared to take on the learning of older persons. U3A facilitators hail from all kinds of backgrounds, and most have no particular pedagogical experience, let alone any specialist degrees in older adult learning. They are unaware of the difference between different learning processes and find it difficult to empathise with the life story of U3A members. Indeed, the problem of preparing specialists to serve older adults exists in many spheres of life, and education is not an exception in Russia. There are very few universities teaching geriatrics and gerontology programmes, and there are none who specialise in geragogy. However, some support by non-governmental organisations is available. For example, whilst Nevsky Angel in St. Petersburg launched a programme for the preparation of teaching volunteers (Nevsky Angel, n.d.), the *Znaniye* branch in Oryol holds regular seminars for U3A educators (Oryol Oblast Branch of Association Znaniye of Russia, n.d.). Finally, developing curricula and teaching materials targeting older adults' needs and preferences are another challenge. Currently, many institutions offering courses for older adults simply copy curricula, syllabi, and materials from study programmes aimed at young students.

It is, however, encouraging that a positive effect on the well-being of those who attend different courses is evident. According to research conducted at the Tuymazy U3A, the majority of learners enjoy classes and look forward to attending lectures (Minnigaleeva, 2013). Moreover, the research in Mariy State University demonstrated that 84% of older students think that learning helps socialisation in older persons, and 81% are satisfied with the content of the study programmes on offer (Domracheva et al., 2016). Over half—52% of the students—also noted that due to participation in the U3A, their quality of life had improved. Respondents mentioned that they had been experiencing personal and spiritual growth and that the U3A enabled them to broaden their lives. They had found friends and been able to improve their memory and health. In a programme dealing with psychological issues, all participants reported that they had gained new life goals following the completion of this study unit (Potrikeyeva & Suprunenko, 2017). Impact on attitudes towards older persons in society is quite difficult to measure, and no research study has addressed this quandary so far. Yet, it is evident that overall publicity on U3As has generated a more positive outlook towards older persons in recent years. Due to the growth of active ageing and learning programmes in Russia, mostly related to U3A activities, there have been more reports in mass media and social networks about how Russians are ageing more actively, successfully, productively, and overall, more positively.

Conclusion and Future Outlook

Older adult learning programmes emerged first in Russia in the 1990s, through the *Znaniye* Association, to enable and empower older persons to adapt to the country's changing economic and political realities. Such learning initiatives were later followed by other programmes organised by non-governmental organisations and governmental social service centres. Currently, older persons in Russia enjoy various forms of the third age learning opportunities offered across the whole country, many of which are supported or directly organised by local or regional authorities. Since one of the targets of the *Federal Strategy for Older Persons* in 2016 was specifically the development of third age learning, one is witnessing more development and growth in the Russian U3A movement. Learning opportunities for older adults have become highly visible in the mass- and social media and are probably the main contributors to active ageing in Russia today. However, the overall level of participation of older persons in both formal and non-formal learning does not exceed 17% for the youngest cohort, and as the majority of older learners are women who seek U3As as a means of improved socialising with same aged peers, it means that both older males and older persons with physical and cognitive challenges are not participating in third age learning (Bondarenko, 2017). It is positive that the majority of older persons are satisfied with the activities within their respective U3As, but measuring the centres' true effectiveness and impact remains problematic, in many cases due to funding issues. Other problems requiring attention range from ensuring that facilitators at U3As are well qualified to facilitate older adult learning classes to the designing of curricula which are of interest to both healthy and vulnerable older persons. The situation may change, however, considering the very recent government decision to raise the pensionable age by the year 2028, as it may trigger changes in curricula, and available opportunities and forms of education. Whilst today's U3A learning programmes are still mostly informal and focus mainly on leisure interests, socialising and engaging seniors in an active life style, this may change towards more professional training and career development in the foreseeable future.

References

Afanas'eva, L. S., & Savina T. V. (2015). Osobennosti prepodavanija anglijskogo jazyka pozhilym ljudjam v ramkah koncepcii lifelong learning (obuchenie v techenie vsej zhizni). *Russkij i inostrannye jazyki i metodika ih prepodavanija, 4*, 140–145.

Authorities of the Jewish Autonomous Region. (2018). *Upravlenie po vnutrennej politike Evrejskoj avtonomnoj oblasti. VEAO otkrylsja pjatyj potok «Universiteta tret"ego vozrasta»*. http://www.eao.ru/gubernator/press-sluzhba-gubernatora-soobshchaet/v-eao-otkrylsya-pyatyy-potok-universiteta-tretego-vozrasta/. Accessed September 11, 2018.

Barsukov, V., & Chekmareva E. (2017). Posledstvija demograficheskogo starenija i resursnyj potencial naselenija «Tret'ego » vozrasta. *Problems of the territory development, 3*(89), 92–108.

Belov, S. V. (2014). Osobennosti razvitija informacionnoj kompetentnosti u ljudej pozhilogo vozrasta. *Pedagogical Education, 9*, 170–172.

Bondarenko, N. V. (2017). Stanovlenie v Rossii nepreryvnogo obrazovanija: Analiz na osnove rezul'tatov obshherossijskih oprosov vzroslogo naselenija strany. *Monitoring of Education Economics., 5*(104), 1–28.

Computer Academy for Pensioners. (n.d.). *Computer academy for pensioners.* www.pc-pensioneru. ru. Accessed October 20, 2018.

Dmitrieva, A. V. (2018). Social 'noe vkljuchenie pozhilyh: prodlenie zanjatosti ili "prodvinutyj" dosug? *The Journal of Social Policy Studies., 1,* 37–50.

Domracheva, S. A., Morova, N. S., Makarova, O. A., & Mal'ceva, E. V. (2016). Monitoring obrazovatel"nyh potrebnostej lic pozhilogo vozrasta v sisteme nepreyvnogo obrazovanija. In N. S. Morozova & N. A. Biryukova (Eds.), *Monitoring of development of educational systems: Theory, practice, analytics.* Collection of articles of Mariy State Univerisity (pp. 119–132). Yoshkar-Ola: Mariy State Univerisity.

Formosa, M. (2014). Four decades of Universities of the Third Age: Past, present, and future. *Ageing & Society, 34*(1), 42–66.

Formosa, M. (2016). Malta. In B. Findsen & M. Formosa (Eds.), *International perspectives on older adult education: Research, policies, practices* (pp. 161–272). Cham, Switzerland: Springer.

Government of Russia. (2016). *Ob utverzhdenii Strategii dejstvij v interesah grazhdan starshego pokolenija.* http://government.ru/docs/21692/. Accessed October 20, 2018.

House of Projects. (n.d.). *House of projects.* https://hprojects.ru/. Accessed October 20, 2018.

Levintov, A. (2015). *Silver University.* Moscow: Center for Educational Studies of Moscow Management School.

Minnigaleeva, G. A. (2004). *Social'naja i pedagogicheskaja rabota s pozhilymi ljud'mi* (Social and pedagogical work with older persons). Unpublished Doctoral Dissertation, Moscow State Pedagogical University, Moscow, Russian Federation.

Minnigaleeva, G. A. (2013). Traditions and recent developments in learning in later life in the Russian Federation. *International Journal of Education and Ageing., 3*(1), 63–72.

Moscow Government. (2018). *Career centre.* https://talent.mos.ru/about/news/129681462/. Accessed October 20, 2018.

Moscow Silver University. (n.d). *Educational materials.* https://su.mgpu.ru/materialy/. Accessed September 20, 2018.

National Research University of Informational Technologies, Mechanics and Optics. (n.d.). *Third age university.* http://u3a.ifmo.ru/ Accessed September 10, 2018.

Nevsky Angel. (n.d.). *Charity association Nevsky angel 30 years.* www.kdobru.ru/. Accessed October 20, 2018.

Ovchinnikov, G. A. (n.d.). Obrazovanie i prosveshhenie pozhilyh ljudej v sisteme nepreryvnogo obrazovanija. Obuchenie v techenie vsej zhizni. http://www.znanie.org/Projects/Age3/Krasnoyarsk/Ovchinnikov1.doc. Accessed October 1, 2018.

Potrikeyeva, O. L., & Suprunenko, G. A. (2017). Obrazovanie pozhilyh ljudej kak sredstvo ih social'noj adaptacii. *Discussion, 6*(80), 111–115.

Russian Statistics Office. (2018). *Chislennost' naselenija po otdel'nym vozrastnym gruppam.* http://www.gks.ru/free_doc/new_site/population/demo/progn3.xls. Accessed October 11, 2018.

Sidorchuk, T. A. (2013). Obrazovatel'nye programmy dlja pozhilyh ljudej. *Bulletin of Saratov University., 3,* 289–294.

Silver Age. (n.d.). *Silver age.* http://ano-crst.ru/. Accessed October 10, 2018.

Tolkacheva, Y. (2018). Kak-uchatsya-v-rossii-te-komu-za-50 Kak uchatsja v Rossii te, komu za 50. *Razbiraemsja s vozmozhnostjami nepreryvnogo obrazovanija dlja pozhilyh rossijan.* https://newtonew.com/culture/kak-uchatsya-v-rossii-te-komu-za-50. Accessed October 20, 2018.

Union of Pensioners of Russia for the Republic of Tatarstan. (n.d.[a]). *Years are not a problem for us* http://godanebeda.ru/. Accessed October 20, 2018.

Union of Pensioners of Russia for the Republic of Tatarstan. (n.d.[b]). *University of the Third Age.* http://www.sprrt.ru/soyuz-pensionerov/universitet-tretego-vozrasta/ Accessed October 21, 2018.

Valerianova, S. (2014). Regional'nyj opyt Narodnyj universitet: Vek zhivi - vek uchis'. *Information bulletin 'Rights of the elderly' 1*, 14–18. http://timchenkofoundation.org/activities/assistance/lib/IB-PravoPojilyh_01.pdf. Accessed October 31, 2018.

Wisdom Ripening. (n.d.). *People's university for retirees*. https://moigoda.org/university/. Accessed October 20, 2018.

Znaniye. (n.d.). *Education in the third age*. http://www.znanie.org/Projects/Age3.html. Accessed October 20, 2018.

Gulnara Minnigaleeva received her Ph.D. in Education in 2004 with a dissertation on 'Social and pedagogical work with older persons', following which she was accepted by the Humphrey Institute of Public Policy (University of Minnesota, USA) as a Fulbright scholar. In 2006, during her internship at the Ageing and Life Course Department at the World Health Organization's (WHO) headquarters in Geneva, she became involved in the WHO Global Age-Friendly Cities Project. Subsequently, she took the leadership of the project in her hometown in Russia, and used its recommendations as a basis to create Organization of Retired Persons titled 'Wisdom Ripening'. In addition to being the Chair of 'Wisdom Ripening', she is also presently working as an Associate Professor at the National State Research University Higher School of Economics in Moscow, where she is engaged in teaching, including courses on ageing.

Chapter 10
From University Extension Classrooms to Universities of Experience: The University of the Third Age in Spain

Feliciano Villar

Introduction

Spain is one of the most aged countries in the world. According to the Spanish National Institute of Statistics (2016), in 2016 there were 8,657,705 people aged 65 years and over, representing 18.4% of the total population. Furthermore, the proportion of people aged 65 and over is expected to rise in the coming decades to reach 36.5% of the total population in 2050 (Abellán, Ayala, & Pujol, 2017).

The demographic trends that Spain is experiencing had wide-ranging impacts on all areas of public policy including education and learning. There are two particularly important factors that seem to influence the participation of older people in learning activities: health status and educational level (Formosa, 2014, 2016). As regards health, Spanish older people can expect to enjoy an optimum level of health, free of disabilities or severe chronic illnesses, for a substantial amount of years after retirement. Whilst Spanish men at 65 are expected to spend 50.5% of their remaining lifetime in good health, this percentage falls to 38.5% in the case of women (Abellán et al. 2017). In terms of educational level, there has been a dramatic change in recent decades. For example, in 1996 only 7.3% of people aged 55–64 had attained tertiary education (levels 5–8 according to the 2011 ISCED classification). This figure had doubled by 2006, 10 years later (15.3%), and by 2017 almost 1 in 4 people aged 55–64 had studied at university (23.2%). In contrast, the number of people who had only attained primary or compulsory secondary education (levels 0–2 according to the 2011 ISCED classification) fell from 88.3% in 1996 to 58.3% in 2017 (Eurobarometer, 2017). These changes reflect the revolution in social norms, political openness, and future prospects that the present generations of Spanish older people have experienced in their lifetime. They lived under the Franco dictatorship in their childhood and youth, a regime that severely curtailed social participation

F. Villar (✉)
University of Barcelona, Barcelona, Spain
e-mail: fvillar@ub.edu

© Springer Nature Switzerland AG 2019
M. Formosa (ed.), *The University of the Third Age and Active Ageing*, International Perspectives on Aging 23, https://doi.org/10.1007/978-3-030-21515-6_10

and was dominated by extreme religious values. However, with the arrival of democracy following Franco's death in 1975, and particularly after joining the European Economic Community (afterwards named the European Union), living conditions and socio-economic prospects improved dramatically for a substantial proportion of the Spanish population. Inequalities decreased, and tolerant, secular, and democratic values were rapidly adopted.

Bearing the above context in mind, this chapter is aimed at describing the emergence, consolidation, and future prospects of the Universities of the Third Age (U3A) in Spain. To this end, it first examines the conceptual framework that fostered their appearance and growth. Second, the chapter describes the diversity of the U3A movement in Spain, which is characterised by the coexistence of at least two different approaches to participation and learning. Third, it will review research on the status and impact of U3As. Finally, the chapter concludes by outlining some of the future challenges that the U3A movement will have to address in order to meet the needs of the new generations of Spanish older people.

Conceptual and Historical Background of the U3A in Spain

The U3A movement appeared in Spain some 10 years after it was first established in Toulouse in 1972. This delay was due to the political and economic conditions of the 1970s in Spain, a decade which witnessed the end of the Franco dictatorship and the transition to democracy. The consolidation of democracy in the late 1970s and early 1980s generated an explosion of political, social, and cultural activities at all levels of society, which facilitated the emergence of initiatives aimed at increasing the educational level of people who, due to the economic and social conditions of post-civil war Spanish society, had not had the opportunity to receive a proper (or often, any) formal education. These initiatives, at first mostly focused on literacy, rapidly evolved and formed the germ of the U3A movement, giving rise to the first university extension classrooms (or university classrooms for older people). These changes ran in parallel with the adoption in the 1980s of a new framework for understanding older people and older age, and with the attendant transformation of policies concerning older adults. This paradigm shift, which had profound implications that favoured the appearance of new educational opportunities for older people, is based on the following principles:

- Emphasis on improving quality of life as the goal of every service and programme targeting older people. This principle is included in Article 50 of the Spanish Constitution, approved in 1978, which states that 'public authorities…shall promote their [older people's] well-being through a social service system meeting specific health, housing, culture, and leisure needs'. In this respect, education in later life was deemed a way to improve well-being, create and reinforce new social bonds, favour social inclusion, and promote self-determination and awareness of rights.

- A new view of older people that, far from the traditional images of disability, dependence, and frailty, recognises their capacities, and right and potential to continue developing and growing. This positive approach to ageing challenges the view of education and learning as being exclusively associated with earlier stages of life (Villar, 2005). In contrast, it underlines that learning in later life is not only possible, but is also a key activity that helps to reinforce and stimulate cognitive and social skills, which in turn promote new opportunities for exercising meaningful roles and guaranteeing rights.
- The adoption of the 'active aging' framework as the epitome standard for ageing well (World Health Organization, 2002). This includes participation in social, economic, civic, cultural, and spiritual activities as a fundamental pillar of ageing well and consequently represents a goal that should guide social policy targeting older people. From this point of view, learning in later life is considered both an expression of active ageing and a means to support other active ageing activities (Villar, Serrat, & Celdrán, 2016). This link between active ageing and learning has been recognised in the political sphere. For instance, in the case of Spain the *White Paper on Active Ageing* (Instituto de Mayores y Servicios Sociales, 2011), commissioned by the Spanish government, included a chapter on lifelong learning and training.

The impact of this conceptual framework is reflected in policy development. For instance, 1978 witnessed the creation of the National Institute for Social Services (INSERSO), now called the National Institute for Older People and Social Services (IMSERSO), and initially aiming at the protection of older people and people with dependencies or at risk of social exclusion. This organisation spearheaded the shift in public policies for older people from a paternalistic and care-oriented approach to a social-, health-, and rights-oriented model (Campos, 1996). During this transition, the 1992 first *National Plan for Older Persons* represented a definitive landmark. As well as regulating the provision of health and social services for older people with dependencies or at risk of exclusion, it promoted actions aimed at integrating older people in their community, using culture and leisure as facilitators. Similarly, it promoted older people's associations and the social prominence of older people. In a highly decentralised country such as Spain, social services and education policies depend to a large extent on regional governments; nevertheless, in the 1990s most regions enacted their own plans for older people, adopting similar lines of action. Within this conceptual and political framework, social and community centres for older people flourished in the 1980s and 1990s. These were originally conceived as spaces for socialising and leisure, but learning and teaching programmes were progressively introduced and became one of their most popular activities. In many cases, these educational endeavours formed the germ of university classrooms for older people, providing the public and setting the stage for the expansion of the U3A movement.

The role of Universities was another fundamental factor in the emergence of the U3A in Spain. The year 1983 witnessed the enactment of the Law on University Reform which was aimed at renovating the Spanish university system to align it with the democratic values of the 1978 Spanish Constitution which defined universities as having four main missions: (i) to create, develop, and disseminate science, technical knowledge, and culture; (ii) to train professionals; (iii) to provide scientific support for the cultural, social, and economic development of the country; and (iv) to extend a university culture. In line with the latter two missions, many universities saw lifelong learning and older learning as a natural area for expansion. This would guarantee their presence and influence in society, especially among a segment of the population that did not form their natural market but which was eager to enjoy the prestige of attending a university-sponsored programme. At the same time, programmes for older people were a way to exert corporate social responsibility, particularly for institutions that depended on public budgets (Ortiz-Colón, 2015). Thus, in the 1990s, many universities established their own U3A programmes under different names (university of experience, university for the aged, senior university, etc.) This process was supported by the creation, also in the 1990s, of many regional universities in Spain, a number of which offered a U3A programmes from the outset. Such programmes helped to create strong ties between the new universities and their communities.

As a result of the above activities, the year 2005 witnessed that launch of the *State Association of University Programmes for Older People* (Asociación Estatal de Programas Universitarios Para Mayores (AEPUM). The mission of this umbrella association for university education programmes targeting older people is to promote programmes aimed at satisfying older people's educational needs. At the same time, it aspires to be a hub of information, research, and assessment on education for older people. To fulfil these objectives, it organises activities such as a biannual congress on U3As and an annual research award for studies focused on U3As. At present, around 50 different university programmes are affiliated to the AEPUM, although is estimated that there are more that are not affiliated. In 2016, there were over 52,000 older students enrolled in U3A programmes affiliated to the AEPUM, more than doubling the 15,000 students participating in the AEPUM's U3A programmes in 2005, the year in which the association was created (Instituto de Mayores y Servicios Sociales, 2011). Nevertheless, despite the consolidation and increasing popularity of the U3A and other educational initiatives for older people in Spain, no specific law regulating education in later life has been enacted to date, and the different policy documents on lifelong learning and training in adult life are clearly biased towards labour market issues, viewing training programmes in adulthood as a means to reskill and retrain workers in order to increase their productivity, or if they are unemployed, to increase their chances of finding work (e.g. Ministerio de Educación, 2011).

The Dual Nature of the U3A Movement in Spain

As discussed above, both the State Association movement and universities played a key role in the emergence of the U3A in Spain. Due to this dual influence, U3A programmes in Spain can also be divided into two main models: university extension classrooms or universities of experience (Guirao & Sánchez, 1997). Both have the same explicit goals, namely to improve quality of life and promote the social integration and participation of older people (Alemán, 2013), but their organisation and methodologies are clearly different.

The Bottom-up Approach: University Extension Classrooms

What are known as 'university extension classrooms' date back to the early 1980s and were the first category of U3As to be founded in Spain. They were developed at the initiative of older people themselves and leveraged the rise of associations in the early years of democracy, being particularly popular in the region of Catalonia (Blázquez, 2002). In academic terms, a typical university extension classroom consists of a number of lectures (normally one or two per week, lasting 60–90 min) organised into academic terms. Traditional lecturing is the primary teaching method, and each lecture is independent from another. Thus, they encapsulate a single lesson or issue, which begins and ends in the session, with no relation to the previous or the following lecture. Generally, such lectures do not entail any kind of formal learning evaluation procedure. Consequently, the result is a very diverse collection of lectures whose content ranges from science to the humanities. As well as the lectures, university extension classrooms also programme outdoor activities, such as excursions and cultural visits. These are sometimes linked to the academic content of a specific lecture (e.g. in Barcelona, a lecture on Gaudi's architecture could be complemented by a visit to his main buildings), but are often of an exclusively social and recreational nature.

From an organisational point of view, the main characteristic of university extension classrooms is that the programme is managed by a board elected by the students themselves. The board is responsible for deciding academic questions (i.e. at the beginning of term, the board decides on the number of lectures, the timetable, and the topics to be addressed in each lecture) and financial issues. Students pay a generally modest membership fee, which is used to pay for lecturers and some outdoor activities (others require additional payment, particularly for travelling or meals). Hence, university extension classrooms must generally have at least 30–40 members to be sustainable. If the number of membership applicants exceeds the available resources (i.e. spaces), new applications are restricted, or if possible, a new classroom is created with the same lectures but programmed at different times. University extension classrooms are provided with the help of universities through partnership agreements. However, the supporting role of universities in this model is generally limited to providing teachers (according to the topics selected by the board), methodological advice, and infrastructure support, such as classrooms, computers, or projectors where necessary. Some universities also offer some kind of affiliation to students, so that they can access the library and its resources.

In sum, university extension classrooms offer flexible learning opportunities, combining the academic rigour of lectures taught by university staff with recreational and social activities. Participants can decide which lectures to attend or not and can renew their membership as often as they wish. Each classroom is independent and can be tailored to the needs and wishes of its members. However, the model also presents some drawbacks: there is little room for generational renewal as classroom members and boards tend to stay for many years. From an academic point of view, the unstructured and eclectic selection of topics offered renders it very difficult to explore anything in depth or provide any more than a general overview.

The Top-Down Approach: Universities of Experience

In the 1990s, a second, relatively top-down model of the U3A began to appear in Spain, generally referred to by the title of 'universities of experience'. In contrast to university extension classrooms, whose emergence was linked to grass-root associations, in the case of the universities of experience, it has been the universities themselves as educational institutions that have established these programmes to meet their goals of extending university culture and forging ties with their surrounding community.

On one hand, programmes are generally designed as versions of formal degrees in selected subject areas. Accordingly, classes are organised into courses (typically, one 2-h or two 1.5-h classes a week per course), terms (typically, two to five courses each term), and academic years (typically, three to five academic years to complete a degree). Course content and classes are tailored to older students, which usually means that the level of difficulty is lower and evaluation requirements (e.g. exams) are absent. Consequently, such programmes are far more standardised and organised than university extension classrooms. Although without obtaining any kind of valid certification for the labour market, students acquire detailed knowledge in one specific discipline or professional area. As has been traditional in the Spanish university system, the methodology employed in universities of experience tends to be very academic and lecture-oriented, and the transition to more active learning models with continuous assessment, as promoted by the European Higher Education Area, is particularly difficult in the case of universities of experience. On the other hand, the content, courses, and global design of the programmes are decided by university staff. In contrast to university extension classrooms, the role of older people is often limited to that of students on the programme. Although some universities of experience have a programme advisory board composed of students elected by other students, the real decision-making power of such boards is quite limited. The only admission requirement for universities of experience is age, with 50 or 55 being the usual lower age limit. Students typically pay a small fee (at least compared to that paid by regular students) each term or academic year, and in some cases the university receives public funding (generally from regional education authorities) to cover some of the running costs, thus assuring the sustainability of the programme.

The classes facilitated by university of experience are taught in the same faculties and classrooms as regular degrees, increasing the symbolic value of 'being at university' for older students and facilitating intergenerational exchanges. In fact, this intergenerational component has been increasingly promoted by some universities of experience, which have begun to include regular courses (not just tailored and restricted to older students) in U3A programmes. A generational mix could serve to enrich the learning experience and contribute to the sustainability of the programme, since fewer lecturers are needed. However, it also has its limits. Older and regular students have very different motives and goals and consequently the number of older students on regular courses is often limited. In addition, older students do not sit examinations or complete assignments aimed at assessment. Overall, this model offers older students a more complete university experience than university extension classrooms. Students receive structured classes with temporal and thematic continuity that have been designed by university staff, which implies high-quality teaching and the possibility of learning more complex issues. As disadvantages, students play a more passive role since they participate little in programme design or management, and degree courses come to an end after a given number of academic years, which may leave some students wishing for more. In this respect, it is not unusual for university of experience users to complete more than one degree. University extension classrooms do not present this problem, since their lectures are different from year to year.

Research on U3As

In recent years, and in parallel with the spread and consolidation of the U3A movement across the country, several studies have been conducted to explore its characteristics, determinants, and impacts.

One research strand aimed at ascertaining the profile of U3A students, since only a minority of Spanish older people attends educational programmes. For instance, Villar and Celdrán (2013) found that less than 9% of people aged between 60 and 75 years old participated in non-formal learning activities, including U3A programmes. In a study comparing U3A participants with the general older adult population, Alfageme (2007) found that U3A students were relatively young (very few were over 70 years old) and tended to have a high socio-economic and educational level. The proportion of females was also higher among U3A students than in the general population. Lirio (2008) confirmed this profile and also found that U3A students tended to have a very good self-perception of health and were very active participating in other associative activities. Similarly, Villar, Serrat, and Celdrán (2016) found that involvement in leisure and productive activities was a stronger predictor of participation in educational courses than sociodemographic variables. Perhaps as a result of the different characteristics of the two U3A models, older students attending university extension classrooms seem to differ from their university of experience counterparts, tending to be older and with a lower level of formal education. Each model also requires a different level of involvement, being far greater in terms of days and hours spent per week among university of experience students (Villar, Pinazo, Triadó, Celdrán, & Solé, 2011).

A second research strand that has generated much research interest concerned the motives for enrolling in U3A programmes, with the consensus being that older adults enrol in earning programmes for more expressive and intrinsic reasons than for instrumental ones (Valle-Aparicio, 2014). For instance, Villar, Pinazo, Triadó, Celdrán, & Solé, (2010), Villar, Triadó, Pinazo, Celdrán, & Solé, (2010) found that regardless of the model (whether university extension classrooms or universities of experience) expressive motives such as acquiring knowledge, expanding the mind, or learning for the joy of learning were the most important reasons given for joining a U3A programme. Another important but secondary motive was to access a larger or stronger social network. Other motives, such as to improve family relationships, to escape from loneliness or boredom, or to forget about life problems, were infrequently reported. Similar results have been obtained in quantitative (Villar & Celdrán, 2014) and qualitative (Villar, Pinazo, et al. 2010; Villar, Triadó, et al. 2010) studies. These findings suggest that members' reasons for enrolling in a U3A programme are similar to those that prompt universities to organise these kinds of learning programmes; motives that according to Alfageme and Cabedo (2005) correspond to the social integration and personal fulfilment of students. Hence, U3As in Spain generally assume that students do not need or desire to participate for instrumental reasons and that the U3A has no role in promoting competences that may enable students to manage their life better, contribute more to their community, or become more involved in society.

A third line of research was the impact on students of attending U3A programmes. The acquisition of knowledge is an obvious and important outcome of participation (Alfageme, 2007), but studies also indicated that U3As may have a significant impact on participants' quality of life. Dimensions such as well-being, personal growth, social support, and even perceived health have been found to benefit from participation in U3As (Orte, March, & Vives, 2007; Pérez-Albéniz, Pascual, Navarro, & Lucas-Molina, 2015). Such benefits seem to be particularly high among students with a lower educational level and among those who are more involved in the programmes (Villar, Pinazo, et al. 2010; Villar, Triadó, et al. 2010). Montoro-Rodriguez and Pinazo (2005) have even suggested that some benefits, such as social ones (e.g. social competences or size of and satisfaction with social networks), increase with years of participation. Finally, research has also explored members' evaluation of U3As where results indicated that members tended to express high satisfaction with the different organisational, academic, and teaching aspects of the programmes. However, but when asked, they were also able to identify weak points of the programme in which they participated. The most frequently mentioned aspects in this respect were related to methodology (e.g. sessions too short or too difficult to follow, lack of space for student participation) (Villar et al., 2011).

Future Challenges for U3As

The U3A movement has been a huge success in Spain. In just a few decades, U3As have spread so rapidly that today virtually all Spanish public and private universities offer their own U3A programme. These programmes are viewed very positively by their students, who also perceive significant benefits from participation. However, this success must not lead to complacency. Spanish U3As face future challenges that could compromise their continued growth and significance, particularly the changing profile of future older generations in Spain, who will have a much higher educational level and will be more empowered, and therefore more demanding as regards what the U3A is able to offer. Other challenges include:

- Once the U3A has been consolidated, efforts and resources should be aimed at improving and guaranteeing programme quality. Establishing a series of quality standards, investing in teacher training, and increasing the implementation of learning assessment methods are positive steps in this direction (Palmero & Eguizabal, 2008).
- Progress should be made on adapting U3As to the European Higher Education Space, which stresses student-centred learning methodologies, favours continuous evaluation, and supports national and international student exchange programmes.
- In a similar vein, the participation of older students should be promoted in all the processes of U3A functioning. Not only should the programmes adopt new participatory methodologies, but new channels should also be implemented that enable older students to participate in the design and evaluation of U3A programmes.
- U3As should try to reach beyond their traditional audience of well-educated, urban, affluent, and healthy older students, to include other profiles such as immigrants, people with disabilities, and the older rural population, who are virtually absent from U3As. Technology and distance education could play a key role in meeting this challenge.
- The present focus on expressive-oriented programmes guided almost exclusively by the motives of learning-for-the-sake-of-learning, and social integration may not meet the learning needs of future older generations. These programmes may need to include other, more instrumentally oriented goals that enable older people to learn skills and acquire competences that are of practical use after the programme ends, increasing and guiding their capacity to contribute to their communities, the organisations in which they participate, and their families (Villar & Celdrán, 2012).
- In the research arena, studies conducted to date on U3As have been too basic, mainly focusing on cross-sectional analyses of self-perceptions collected in small and intentional samples. Research needs to employ more sophisticated methods (e.g. paired groups, longitudinal designs) and include a wider range of subjective and objective indicators. The inclusion or older people in the design, analysis, and discussion of research would also be a welcome move.

Facing these challenges would probably be easier if there were some kind of legislation recognising the U3A and in general, the individual and societal value of learning and training in later life. At present, older people are considered in laws on dependency, pensions, and health provision, underlining risks and their vulnerable position in society, but we also need to recognise their potential to grow and to contribute to others. In our view, supporting training in later life paves the way to viewing older people as a resource, not just as a burden.

References

Abellán, A., Ayala, A., & Pujol, R. (2017). Un perfil de las personas mayores en España, 2017. Indicadores estadísticos básicos. *Informes Envejecimiento en Red 15*. http://envejecimiento.csic. es/documentos/documentos/enred-indicadoresbasicos17.pdf. Accessed February 28, 2018.

Alemán, C. (2013). Políticas públicas para mayores. *Gestión y Análisis de Políticas Públicas 9*. http://www.redalyc.org/articulo.oa?id=281528255001. Accessed February 28, 2018.

Alfageme, A. (2007). The clients and functions of Spanish university programmes for older people: A sociological analysis. *Ageing and Society, 27*(3), 343–361.

Alfageme, A., & Cabedo, S. (2005). Los programas universitarios para mayores. In S. Pinazo & M. Sánchez (Eds.), *Gerontología: Actualización, innovación y propuestas* (pp. 367–389). Madrid: Pearson Educación.

Blázquez, F. (2002). Los mayores, nuevos alumnos de la Universidad. *Revista Interuniversitaria de Formación del Profesorado, 45*, 89–105.

Campos, B. (1996). La construccion de una politica social de vejez en España: del franquismo a la normalizacion democrática. *Revista Española de Investigaciones Sociológicas, 76*, 239–263.

Eurobarometer. (2017). *Population by educational attainment level, sex and age*. http://appsso. eurostat.ec.europa.eu/nui/submitViewTableAction.do. Accessed February 28, 2018.

Formosa, M. (2014). Four decades of universities of the third age: Past, present, and future. *Ageing and Society, 34*(1), 42–66.

Formosa, M. (2016). Malta. In B. Findsen & M. Formosa (Eds.), *International perspectives on older adult education: Research, policies, practices* (pp. 161–272). Cham, Switzerland: Springer.

Guirao, M., & Sánchez, M. (1997). Los programmeas universitarios para mayores en España. In A. Lemieux (Ed.), *Los programmeas universitarios para mayores: Enseñanza e investigación* (pp. 145–153). Madrid: IMSERSO.

Instituto de Mayores Servicios Sociales. (2011). *Libro Blanco del envejecimiento activo*. Madrid: IMSERSO. http://www.imserso.es/InterPresent1/groups/imserso/documents/binario/ 8088_8089libroblancoenv.pdf. Accessed February 28, 2018.

Lirio, J. (2008). *La gerontología educativa en España*. Unpublished doctoral dissertation. http:// eprints.ucm.es/8315. Accessed February 28, 2018.

Ministerio de Educación. (2011). *El aprendizaje permanente en España*. Madrid: Ministerio de Educación. https://sede.educacion.gob.es/publiventa/ImageServlet?img=E-14856.jpg. Accessed February 28, 2018.

Montoro-Rodriguez, J., & Pinazo, S. (2005). Evaluating social integration and psychological outcomes for older adults enrolled at a University Intergenerational Programme. *Journal of Intergenerational Relations, 3*(3), 65–81.

Orte, C., March, M. X., & Vives, M. (2007). Social support, quality of life, and university programmes for seniors. *Educational Gerontology, 33*(11), 995–1013.

Ortiz-Colón, A. M. (2015). Los Programmeas Universitarios de Personas Mayores y el Envejecimiento. *Activo: Formación universitaria, 8*(4), 55–62.

Palmero, C., & Eguizábal, A. (2008). Quality of university programmes for older people in Spain: Innovations, tendencies, and ethics in European higher education. *Educational Gerontology, 34*(4), 328–354.

Pérez-Albéniz, A., Pascual, A. I., Navarro, M. C., & Lucas-Molina, B. (2015). Más allá del conocimiento El impacto de un programmea educativo universitario para mayores. *Aula Abierta, 43*(1), 54–60.

Valle-Aparicio, J. E. (2014). Educación permanente: Los programas universitarios para mayores en España como respuesta a una nueva realidad social. *Revista de la Educación Superior, 43*(171), 117–138.

Villar, F. (2005). Educación en la vejez: Hacia la definición de un nuevo ámbito para la Psicología de la Educación. *Infancia y Aprendizaje, 28*(1), 63–79.

Villar, F., & Celdrán, M. (2012). Generativity in older age: A challenge for universities of the third age (U3A). *Educational Gerontology, 38*(10), 666–677.

Villar, F., & Celdrán, M. (2013). Learning in later life: Participation in formal, non-formal and informal activities in a nationally representative Spanish sample. *European Journal of Ageing, 10*(2), 135–144.

Villar, F., & Celdrán, M. (2014). The involvement of Spanish older people in non-degree educational programmes: Reasons for and barriers to participation. *Journal of Aging and Social Policy, 26*(4), 370–385.

Villar, F., Pinazo, S., Triadó, C., Celdrán, M., & Solé, C. (2010). Older people's university students in Spain: A comparison of motives and benefits between two models. *Ageing and Society, 30*(8), 1357–1372.

Villar, F., Pinazo, S., Triadó, C., Celdrán, M., & Solé, C. (2011). How students evaluate university programmemes for older people: A comparison between two models in Spain. *Journal of Aging Studie, 25*, 118–125.

Villar, F., Serrat, R., & Celdrán, M. (2016). Participation of Spanish older people in educational courses: The role of sociodemographic and active aging factors. *European Journal of Social Policy, 26*(5), 417–427.

Villar, F., Triadó, C., Pinazo, S., Celdrán, M., & Solé, C. (2010). Reasons for older adult participation in university programmes in Spain. *Educational Gerontology, 36*(3), 244–259.

World Health Organization. (2002). *Active ageing: A policy framework*. Geneva: World Health Organization.

Feliciano Villar holds a Doctorate in psychology and is currently a Professor at the Department of Cognition, Development and Educational Psychology, University of Barcelona. His research interests include the study of determinants, personal and social consequences, and meanings associated with active ageing lifestyles. As a result of his research activity, he has authored more than 50 papers in 'Journal Citation Reports' journals and has been a Visiting Scholar in more than 10 international higher education institutions, including the University of Oxford (UK), Wilfred Laurier University (Canada), University of Twente (the Netherlands), and the Autonomous National University of Mexico (Mexico).

Chapter 11
Sweden's Senior University: *Bildung* and Fellowship

Cecilia Bjursell

> *I hope that you have understood how much Senior University has meant to me!*
>
> Swedish Senior University student (2018).

Introduction

In Sweden, the Senior University consists of a number of non-profit, volunteer associations, which offer courses and activities to their members.[1] Most associations have a minimum age limit of 55 or even older, and hence, the members can be considered as being in the 'third age' phase of the life course which follows 'middle age'. When one asks participants in Senior Universities why they are involved in the association's activities, two main points are generally raised. First, that they wish to further their education by acquiring new knowledge, and secondly, that they aspire to be part of a social community with other, like-minded and same-aged peers. The number of Senior Universities in Sweden is on the increase, as is the overall number of participants. Currently, there are 34 Senior Universities across the country. They are organised as associations and are formally linked to the Swedish *Folkuniversitet* system, one of ten educational associations that exist in the Swedish *Folkbildning* organisation (a 'general level' education structure for adults). These Senior Universities have a total of 25,000 members. Since participation in the activities is voluntary, when one considers the increasing number of participants it follows that those individuals who enrol do so because they expect to benefit from their membership. The Senior University is seen as something positive for those who are active in the movement,

---

[1] The term *Senior University*, also known as the 'Pensioners' University', is the name used by most associations, and thus, will be used throughout this chapter.

C. Bjursell (✉)
School of Education and Communication, Jönköping University, Jonkoping, Sweden
e-mail: cecilia.bjursell@ju.se

© Springer Nature Switzerland AG 2019
M. Formosa (ed.), *The University of the Third Age and Active Ageing*, International Perspectives on Aging 23, https://doi.org/10.1007/978-3-030-21515-6_11

but there are several different reasons why people attend. This chapter provides an overview of the development of the Swedish Senior University system and explores why individuals choose to participate in the organised learning activities.

Senior University Origins (Uppsala, 1979)

The Swedish movement is based on international ideas about older adult learning at university premises for individuals in the third age. During the 1970s, Anne Walder learnt about an ongoing educational initiative in France and this aroused her curiosity about the possible educational opportunities for retirees. In her own words,

> About 1975, I happened to read in a French educational magazine about a newly-established "Université du Troisieme Age" by a Professor Pierre Vellas of Toulouse. I was working in Haparanda at the time and my own retirement was on my mind. I found his ideas interesting and I wondered whether others, or people in other countries, had the same idea. In 1977, I won a scholarship which allowed me to visit a number of schools in the USA. From Harvard University, Cambridge, (Mass.), I received documents about a newly-established "Institute for Learning in Retirement" along with a recommendation that I also contact a similar organisation in New York.
>
> Walder (1989): 15

Walder travelled to several locations to explore and investigate how educational opportunities were organised for retirees. On her return to Sweden, the idea remained with her and she contacted Olle Alexandersson, then Vice Chancellor of Kursverksamheten (now known as 'Folkuniversitetet'), in Uppsala. In June 1979, he convened with a group of academics who were associated with Uppsala University—and inspired by the logistic techniques already used in France, England, and the USA—they made plans to establish the first University of the Third Age (U3A) in Sweden. On 9 October, 1979, the Uppsala Pensioners University was founded and, in 1985, included 300 members. Its affiliation with the international movement was established with the association's membership of the international organisation, the *Association Internationale des Universités du Troisième Âge* (AIUTA). In the almost 35 years that have passed since then, more than 30 new associations have been established. Uppsala Senior University, like many other associations, is not directly subsumed by its parent university. Instead, it is organised as an association which is linked to *Folkuniversitetet*, one of the largest Swedish study associations.

The Organisation of the Swedish Senior Universities

Senior Universities are associations which are managed by pensioners, for pensioners. However, what distinguishes Senior Universities from other study associations is their links to higher education. The courses on offer are not structured as credit-awarding, formal university courses, but are built around informal learning in communities of interest. Consequently, the courses offered by the senior universities are similar to the study circle courses that are offered in the popular/liberal adult education movement in general.

Adult education, including 'folkbildning', has been available in Sweden since the second half of the 1800s and is an essential part of the Swedish system for lifelong learning. *Folkbildning* is a general term, which refers to the learning that takes place in study associations and folk high schools, as well as to a shared value system. This value system is based on the assumption that education should provide everyone in society the opportunity to become aware of the world we live in and be able to act within this world. *Folkbildning* provides support for personal development and allows for individuals to become aware of their role as members of society. The central issues to the *folkbildning* movement are that it is a free and voluntary activity, which entails that it is the participants themselves who choose what they wish to participate in: 'Folkbildning addresses every person's life-long right to freely search for knowledge' (Folkuniversitet, 2018—online source). The direction and intent of *folkbildning* are expressed in terms of five principles:

- Learning is related to the entire situation in life of the individual. It is a holistic view of people and knowledge, of the enlightenment process as a lifelong journey that leads to personal growth.
- Knowledge and *bildung* have an intrinsic value. *Bildung* refers to a process of personal and cultural maturation through education, which emphasises community, philosophy and art. This contrasts with a purely instrumental view of knowledge, in which knowledge is obtained and communicated for a specific purpose.
- *Folkbildning* is free and voluntary. People voluntarily participate in learning processes that involve social interaction, cooperation, discussion, and reflection—voluntary interactions that benefit people and communities.
- Participants are active in the creation process. Self-organisation, participatory influence, and democracy are practised in activities based on the belief in the human capacity for taking responsibility and in defence of human equality.
- Community engagement is stimulated and channelled through *folkbildning* being rooted in popular social movements and associations, but also through the ability to meet new needs and tackle new issues that engage people in the community, locally and globally, in a flexible and unconventional way.

Currently, there are 156 folk high schools and 10 study associations in Sweden. They are funded by the state as they operate within the policy framework set out by the government. One study association, *Folkuniversitetet*, cooperates extensively with the Senior Universities. What they have in common is that they are both not linked to any political party, labour union, or religious affiliation, but they are linked to higher education organisations. While *folkbildning* is broadly aimed at the whole population, people aged 65-plus actually constitute a large part of the total number of participants in study circles in Sweden (Andersson, Bernerstedt, Forsmark, Rydenstam, & Åberg, 2014). Table 11.1 provides a summary of all of the *folkbildning* activities for 2016 and shows that the oldest age group represented the highest proportion of participants in study circle activities (37%), other *folkbildning* activities (26%), and as a proportion

Table 11.1 The proportion of participants for each age group and the activity type, the total number of participants in study circles and other *folkbildning* activities compared with the national population (2016)

Age group	<12	13–19	20–24	25–44	45–64	65+	Total
Study circle (%)	0	8	6	24	25	37	100
Other *folkbildning* activities (%)	16	9	4	20	25	26	100
Proportion of total participants (%)	7	8	5	22	25	32	100
Number of participants	74,291	75,498	50,233	208,474	228,268	290,899	927,663
Proportion of the national population (%)	15	7	6	26	25	20	100

Source Swedish National Council of Adult Education (2017)

Table 11.2 Participants in Senior University according to geographical region and activity (2016)

Geographical	Study circle		Other *folkbildning* activity		Culture programme	
Region	Hours	Participants	Hours	Participants	Number of activities	Public
East	9477	5728	1044	2264	623	75,325
Mid	5289	3838	170	287	346	51,051
West	5060	2901	66	60	46	3825
South	3379	2188	291	399	65	6020
North	336	173	0	0	212	13,906
Total	23,541	14,828	1571	3010	1292	150,127

Source Folkuniversitetet (2018)

of the total number of participants (32%). In total, 290,899 people from the age group 65 and over participated in *folkbildning* in 2016 (Swedish National Council of Adult Education, 2017).

Similar to many other developed countries, Sweden's population has experienced an increasing number and percentage of older and healthier citizens. In 2015, almost every fifth person was over the age of 65, and it is estimated that in 2060 25% of the population of Sweden will be over the age of 65 (Statistics Sweden, 2016). These statistics are reflected in the increasing interest in Senior Universities, as shown by the increase in the number of associations and their enrolment numbers. Since each Senior University is its own association, there exists no statistics or information about their activities except the figures provided about the level of activity by the participants. Table 11.2 shows the level of participation in terms of hours and the number of individuals active in study circles and other activities, as well as the number of cultural activities and the public's attendance to these activities.

Study circles are non-formal groups of individuals who meet together with the purpose of enriching their knowledge of a subject or specific question. One criterion is that such a group must consist of at least three people who get together on at least three occasions for a minimum total of 9 h of study. Other *folkbildning* activities may include learning that is freer in form than that of a study circle. For example, such activities may include a one-off meeting or a larger number of participants than can be accommodated in a study circle, something which may take place during a themed meeting or on introductory 'have-a-go' days. Culture programmes include activities of various types, for example a music concert or public talk, where the public is invited to the audience, and thus can come into contact with the *folkbildning* movement. The next section continues to examine the reasons why individuals enrol at Senior Universities.

Participants' Voices

In an effort to review the current situation of how participants view their own engage-ment in the Senior University movement, in the spring of 2018 the author sent out a request for information to the different associations across Sweden. Although most associations have policies prohibiting the distribution of questionnaires to members, nevertheless a total of 53 responses, from 38 women and 15 men, were received (4 by ordinary post, the rest by email). These letters contained a range of responses, from short comments of a few lines to one of five pages in length.[2] Both Senior University members and lecturers responded to this request for information, and, in some instances, some held the roles of both participant and lecturer. One person, who reported only a short formal education, stated that Senior University had meant a great deal because it broadened his horizons in an unprecedented manner. Notwith-standing this, it was common for the respondents to hold some form of academic background and careers as, for example engineer, medical doctor, teacher, sociolo-gist, or veterinary, amongst others. The majority of respondents were in the 70–80 age bracket, with the oldest respondent being 93 years old.

An analysis of the contents of the responses uncovered two overarching reasons for participation—namely a desire for *bildung* and fellowship. The participants wished to educate themselves by improving their knowledge of a particular area, either via further development of a subject that is already known to the participant, or by becoming proficient in an area that is new to them. Their participation also offered them a social context, and for many, this dimension was as equally important as the subject content which they endeavoured to study. Indeed, one Senior University

---

[2] The excerpts from the responses have been kept anonymous, with the author communicating only the gender of the author (indicated by 'f' for female and 'm' for male), plus an ascribed unique number to indicate when multiple excerpts or quotations have been made from the same respondent. Moreover, it should be pointed out that the collection of narratives and reflections that were shared with the author did not constitute a systematic investigation but, rather, an exploratory effort to come to an understanding of why older persons are motivated to enrol in Senior Universities.

participant reported that 'the reason to continue with learning for the elderly is to maintain and develop one's knowledge and to be a part of society, and it is social and promotes good health and it gives one a more meaningful life' (f, 18).

In several responses, participants mentioned that it is important to keep one's mind active: 'Stimulation is extremely important for improved well-being and to get input for one's thoughts. Now there is time for reflection' (f, 20). Another respondent claimed that she 'believes that you can keep your brain healthier by learning new things and not least by having contact with other people who you might not normally socialise with—you can't just solve crossword puzzles' (f, 06). A third participant stated that 'being a part of an association or the equivalent is also a very important health factor. It contributes to having a meaningful and rewarding life in one's old age' (f, 05). Participation in Senior University activities was often just one of many possibilities for active ageing for many retirees. For example, the following pensioner's schedule is noteworthy:

> Exercise at the gym 2.5 h, 3 times per week, practice guitar playing, electric bass guitar, French language studies. Will be 72 years old soon. Bought a new caravan last year, obtained a 'green card' for golf in 2016. Travel a lot. (m, 01)

Indeed, many respondents also claimed to have made a conscious decision to continue pursuing an active life after they retire: 'I want to be an active pensioner' (m, 02). The connection between one's health and learning was especially important for participants who are looking for both cognitive and social stimulation.

Bildung *and Fellowship*

From the responses received, it resulted that Senior Universities provided opportunities for both *bildung* and fellowship for its members, as there are several directions that the participants' learning can take. Some participants aspired to be active with advanced studies where they continue to study an area which they have already mastered. Others, however, wished to learn about a subject which they did not previously have opportunity to study. One aspect of further study after retirement is the common consensus that it is important to keep one's brain active. For some participants, the study circle's open format does not suit them and, consequently, they enrol at university instead. Finally, there are many people who view Senior University as an opportunity to engage in mature debates and discussion in diverse areas so as to satisfy their curiosity and to deal with existential questions in open communication. For these participants, being willing to enter into a dialogue with other people, who also possess a great deal of life-experience, provides the conditions where one can speak about things in a way which is perhaps relatively impossible with younger people, who find themselves in earlier stages of the life course. Indeed, conversations with others allow one to broaden one's horizons and one's sense of enlightenment about life in general, not just about specific facts.

The social dimension of participating in Senior University activities is not just about meeting other people and preventing social isolation. By keeping oneself up to date in different fields, one is also able to follow social developments and one can feel that one can still contribute to society as an active citizen. Meeting together, around a common area of interest, can offer greater opportunities for learning since one is able to hear of other people's experiences and their knowledge, something with is considered positive. Social fellowship is thus important, even contributing to the intellectual stimulus experienced by participants. Belonging to a group where the participants share certain characteristics can further the social interaction that takes place. The similarities shared across the participants created a spirit of acceptance and security. Being part of an environment solely made up of older participants can also support the interaction between the participants since they share the same pace of living and have empathy for certain ailments, for example, having auditory challenges. The notion of status may also play a role in being able to identify oneself with the others in the group as a reflection of one's own identity. The search for a context or environment which is characterised by a spirit of acceptance, or where one can maintain one's earlier status in life and identity, are key manifestations of different aspects of the ageing process.

The Study Circle Format

Senior Universities publish course catalogues which are popular reading because they contain so many interesting courses. Despite the fact that they are called *courses* in the catalogue, the presentation of the material is in the form of a 'study circle'. A study circle is based on the active and equal participation of the 'course leader' as well as the other study circle participants. The idea behind this arrangement is that learning takes place at the intersection of other people's knowledge and experiences:

> I am active and happily take the initiative. I am most comfortable in circles where every member has a voice in the discussions, less so when an 'omniscient' leader speaks. People often call the circles *courses*, but if they are thought of as courses, it gives the wrong idea. Naturally the leader should possess a certain amount of fundamental knowledge, but all of the participants must be active and enrich the meetings with their points of view. (f, 53—emphasis in original)

The free and open format of the study circle, which is aimed at the development of each individual, is appealing to all ages, but especially older persons. One Senior University participant reported that the voluntary aspect of one's participation actually contributes to increased learning: 'It is easier to learn when it is fun and there is no compulsion [to learn]. You choose the subjects that you are interested in' (f, 03). Another participant, who was the chair of UNESCO's Institute for Lifelong Learning, claimed that even during his tenure he became tired of the emphasis placed on the utilitarian perspectives on education:

> I was then, and remain so, so tired by the idea that all education should be *useful* in terms of the labour market…the idea that as a pensioner, when less demands were made of you, you

wished to learn something because it might be cool or interesting was seen as something 'inappropriate'. (m, 04—emphasis in original)

There is no doubt that if the meetings become too similar to a traditional classroom environment, then one is lacking an important dimension of the intended format of the activity—namely, that the participants should not be just fed information, but, rather, they should themselves participate in the development of the course and feel that they are contributing to the whole. Citing the response of one male participant,

In our case, I don't know how important the notion of 'learning' is. We do learn, so I feel, by what we do. We seldom speak of 'learning' [per se]. Vocabulary training is something which I see as a personal responsibility, and nothing that is 'checked up on' – but, of course, words and the choice of words are discussed. For me, just suitable texts can be challenge enough. I would really like to consult with colleagues about the same assignment. The management from the governing body (in Senior University) is non-existent; there is encouragement and also trust, which I know is typical of a secure, established organisation. (m, 28)

Discussions about common areas of interest are especially fundamental to book circles—also known as *reading circles*—which is a rather popular activity. Although reading can be done alone, the circle can be used to start off a book and then more thoroughly examine that which was read since one last met:

After a brave but futile attempt at starting reading circles in my home town, following my husband's advice I enrolled on a Senior University program in a different town. I can take the train using my train pass and stay over at my brother's place. The first circle addressed old age and the two following circles were on Umberto Eco's *The name of the rose* and the short stories of Franz Kafka. One circle discussed poetry. (f, 25)

The most popular areas of interest, as mentioned in the participants' responses, included art history, music history, history, high quality films, cultural trips, field trips, jazz, blues, poetry, drama, gender studies, literature, philosophy, psychology, the history of philosophy, photography, the use of smartphones, biology field trips, English conversation, German, Greek loan words, personal development, gardening, art, lyrical poetry, the theatre, opera, bird watching field trips, flower field trips, the history of literature, watercolour painting, creative writing, botany, human health, quality of life, ethics, nature, local history, culture, Spanish, foreign language conversation, reading circles, medicine, religion, ancient civilisations, churches, concerts, social issues, monarchies and republics, Photoshop, and cultural trips to other countries.

Promoting Lifelong Learning

The previous sections focused on the Senior Universities that represent the University of the Third Age movement (U3A) in Sweden. Several older persons participating at Senior Universities mentioned how engaging in learning pursuits functioned to improve their well-being. This is hardly surprising considering that the ability to learn is a lifelong capability that remains embedded in human beings even following the onset of later life (Andersson & Tøsse, 2013). Furthermore, the social engagement has positive effects on the individuals' quality of life. Current age demographics, which include an increasing proportion of senior citizens with unprecedented positive levels of physical capital, have led to a situation which is historically unique, with current older persons being more mobile and *compos mentis* when compared to yesteryear peers. At the same time, society is undergoing rapid change, which demands the development of new knowledge and skills that in turn, warrants citizens to continually upgrade their knowledge. When individuals participate in creative learning activities, this does not only promote increased well-being at that moment in time, it also lays the foundation and preparedness for the individual to develop in pace with the rest of society. As Formosa argued,

> Research studies about the benefits of creativity work for older adults suggest that there are multiple positive effects. Findings imply that creative activities can improve problem-solving ability, self-esteem, coping skills, anxiety, life satisfaction, and depressive symptoms. Moreover, creative interventions have generally elicited positive anecdotal feedback from older adult participants and stimulated their involvement and responsiveness...creative activity contributes to successful aging by encouraging development of problem-solving skills that translate into a practical creativity in older adults' daily living. They also found that creativity fostered a sense of competence, purpose, and emotional growth. Physiological benefits have also been recorded. When the brain engages in creative work, it alerts the parasympathetic nervous system; heart rate and breathing slow down, blood pressure decreases, blood circulation to the intestines increases, and the body shifts into relaxation.
>
> Formosa (2013: 81)

Similar to the *folkbildning* movement, Senior Universities offer educational activities that are based on study circles, field trips, and public lectures. A key aspect of the *folkbildning* movement is that participation is voluntary. This is key to understanding why pensioners choose to engage in a number of diverse learning engagements (Bjursell, Bergeling, Bengtsson Sandberg, Hultman, & Ebbesson, 2014). Senior Universities allow a high degree of flexibility with respect to dividing one's time in maintaining past identities and constructing new ones. Indeed, study circles provide older learners with opportunities for both social and personal development in an equitable and democratic space (Andersson & Tøsse, 2013). Although study circles which are similar in format to traditional classroom teaching practices do exist (Nordzell, 2011), membership in a study circle is never just about aquiring new knowledge, but entailing being part of a creative learning context where individuals can free themselves of old approaches and preconceptions (Nordvall, 2002). Senior Universities, and *folkbildning* in particular, provide a democratic and emancipatory arena for adult learning (Bjerkaker, 2014), acting as a catalyst for the development of the older individual as a member of society. In addition to instantiating an arena for learning, the study circle model's advantage (and success) is based on emphasising the importance of (i) one's sense of personal responsibility and (ii) the social dimension inherent to meaningful learning (Bjursell, 2017). These observations are reflected in the responses provided by the respondents in the previous section. Yet, what differentiates Senior Universities from *folkbildning* is the demand that the content of these activities be of a higher academic level, and hence, resonating more with the Francophone U3A model than the Anglophone one.

When it comes to older adults learning, there is the mistaken assumption that older persons constitute a homogeneous group. Instead, the older a cohort becomes, the more differentiated it becomes and within each generation, each individual has their unique combination of characteristics, expectations, and abilities (Formosa & Higgs, 2015; Higgs & Formosa, 2015). However, as the case with many U3As in other countries, one general commonality amongst Senior Universities' participants is their middle-class and professional background. Many of the respondents in this study held such a background—thus, confirming again that level of educational attainment is perhaps the most significant factor associated with lifelong learning in post-retirement (Bjursell, Nystedt, Björklund, & Sternäng, 2017). In other words, the lengthier of one's level of educational attainment, the higher the probability that he/she will enrol in a Senior University. Citing Formosa again,

> This occurs because older adults who have experienced post-secondary education, and have advanced qualifications and skills, are already convinced of the joy of learning so that their motivation to enrol in U3As is very strong. To middle-class elders, joining means going back to an arena in which they feel confident and self-assured of its outcome and development. On the other hand, working-class elders are apprehensive to join an organisation with such a 'heavy' class baggage in its title. Moreover, the liberal-arts curriculum promoted by most U3As is perceived as alien by working-class elders, who tend to experience 'at-risk-poverty' lifestyles, and are more interested in practical knowledge related to lifelong work practices.
>
> Formosa (2014: 53)

This is not, however, the same as saying that there are no exceptions to this trend. For instance, Golding's (2015) research on the Men's Shed movement found that although older, poorly educated, men are generally reluctant to participate in formal educational activities, an exception arises when such men form part of a group of like-minded individuals in an informal learning situation. Golding (ibid.) demonstrated that if learning meets individuals' life histories and habitus, the learning initiative has the potential to meet the goals of a 'wide-participation' agenda.

Regardless of whether the study group consists of working-class elders or people from an academic background, there is a risk for stagnation instead of development in like-minded groups. A homogeneous group might reinforce pre-existing attitudes and approaches, instead of challenging the participants' viewpoint and broadening their perspectives as they consider the complex nature of society. A homogeneous group is, however, not necessarily a disadvantage. Similarities between the participants create acceptance and a sense of security, and can, in certain cases, be a deciding factor for whether an individual will join the group or not. Since Senior Universities mainly succeed in attracting persons who aspire an academic ambition for their leisure, policy makers in Sweden should consider how to support other individuals who are non-typical members in lifelong learning. They should also consider how to provide courses for students at a later stage of their life cycle in various formats and contexts since 'such provisions will have a vital role in keeping retired people in touch with their social environment' (European Commission, 2006: 8). Although some groups might need stronger support, self-organised associations like the Senior Universities still have a role to play. In the words of a participant: 'Finally, I can just say that it is important that associations like Senior University exists in our society' (m, 02).

References

Andersson, E., Bernerstedt, M., Forsmark, J., Rydenstam, K., & Åberg, P. (2014). *Cirkeldeltagare efter 65: Livskvalitet och aktivt medborgarskap. Folkbildningsrådet utvärderar, 2014*. Stockholm: Folkbildningsrådet.

Andersson, E., & Tøsse, S. (2013). Tid för studier efter sextiofem. In A. Fejes (Ed.), *Lärandets mångfald: Om vuxenpedagogik och folkbildning*. Studentlitteratur: Lund.

Bjerkaker, S. (2014). Changing communities. The study circle—For learning and democracy. *Procedia—Social and Behavioral Sciences, 142*, 260–267.

Bjursell, C., Nystedt, P., Björklund, A., & Sternäng, O. (2017). Education level explains participation in work and education later in life. *Educational Gerontology, 43*(10), 511–521.

Bjursell, C. (2017). *Äldre medarbetares kompetensutveckling*. In: Helene Ahl, Ingela Bergmo Prvulovic & Karin Kilhammar (Eds.), HR: Att ta tillvara mänskliga resurser (pp. 177–190). Lund: Studentlitteratur.

Bjursell, C., Bergeling, I., Bengtsson Sandberg, K., Hultman, S., & Ebbesson, S. (2014). *Ett aktivt åldrande: pensionärers syn på arbete och lärande*. Visby: Nomen/Books on demand.

European Commission, (2006). *Adult learning: It is never too late to learn*. Brussels: European Commission.

Folkuniversitetet. (2018). *Welcome to Folkuniversitetet*. http://www.folkuniversitetet.se/Skolor/Senioruniversitet/. Accessed April 3, 2018.

Formosa, M. (2013). Creativity in later life: Possibilities for personal empowerment. In A. Hansen, S. Kling & Ś. Strami González (eds.), *Creativity, lifelong learning and the ageing population* (pp. 78-91). Ostersund, Sweden: Jamtli Förlag.

Formosa, M. (2014). Four decades of Universities of the third age: Past, present, future. *Ageing and Society, 34*(1), 42–66.

Formosa, M., & Higgs, P. (2015). Introduction. In M. Formosa & P. Higgs (Eds.), *Social class in later life: Power, identity and lifestyle* (pp. 1–14). Bristol: The Policy Press.

Golding, B. (Ed.). (2015). *The men's shed movement: The company of men*. Champaign: Common Ground Publishing.

Higgs, P., & Formosa, M. (2015). The changing significance of class in later life. In M. Formosa & P. Higgs (Eds.), *Social class in later life: Power, identity and lifestyle* (pp. 169–181). Bristol: The Policy Press.

Nordvall, H. (2002). Folkbildning som mothegemonisk praktik? *Utbildning och demokrati, 11*(2), 15–32.

Nordzell, A. (2011). *Samtal i studiecirkel: Hur går det till när cirkeldeltagare gör cirkel?*. Stockholm: Stockholms universitets förlag.

Statistics Sweden. (2016). *Hur stor är Sveriges folkmängd 2060?* http://www.sverigeisiffror.scb.se/hitta-statistik/sverige-i-siffror/manniskorna-i-sverige/framtidens-befolkning/. Accessed March 15, 2016.

Swedish National Council of Adult Education. (2017). *Annual report 2016*. Sweden: Swedish National Council of Adult Education.

Walder, A. (1989). *Så började Uppsala Pensionärsuniversitet. I Uppsala Pensionärsuniversitet 1979–1989*. Uppsala: Uppsala pensionärsuniversitet.

Cecilia Bjursell is Associate Professor and Director of Encell, the National Centre for Life-long Learning. In her research, she is interested in organising, learning, metaphors and narrative approaches in different empirical contexts. Her current projects include learning later in life, learning in organisations and education management. As Encell's Director, collaboration with the surrounding society is a central part of her work. She is a member of several boards, councils and networks.

Part III
Asian-Pacific Perspectives

Chapter 12
The University of the Third Age Movement in Australia: From Statewide Networking to Community Engagement

Ainslie Lamb

The 'Third Age' Demographic in Australia

The third age, defined as the time in one's life course for personal achievement and fulfilment beyond child-raising and retirement from the paid workforce (the 'second age'), is a twentieth-century phenomenon, as life expectancy has extended well beyond retirement age for most people in all continents, including Australia. The third age is not a fixed chronological age, although it is often related to the opportunity to qualify for a government sponsored age pension or a superannuation pension. In Australia, that can mean any age between 55 (when superannuation pensions can be accessed) or 65 (when the age pension can be accessed, to be increased to 67 by 2023). By 2016, 15.7% of the Australian total population was aged 65 or older and is projected to increase to 25% by 2050 (Australian Bureau of Statistics, 2018). By 2016, while life expectancy for Australian males had increased to 80.3 years, for Australian females it reached 84.4 years. As a result, the years of 'retirement' for most Australians can span up to 20 years and even beyond. These years can be enhanced through adoption of a lifelong learning ethos, for which the University of the Third Age (U3A) movement provides rich opportunities.

Origins and Organisation of U3A in Australia

The first U3A to be established in Australia was in Melbourne, Victoria, in 1984, which adopted the 'Cambridge' model co-founded by Peter Laslett in Britain. The concept quickly spread to other States and Territories by 1989. Minichiello (1992) described the early days for the Australian U3A movement as follows:

A. Lamb (✉)
U3A Network New South Wales, Sydney, Australia
e-mail: ainslie2518@gmail.com

© Springer Nature Switzerland AG 2019   145
M. Formosa (ed.), *The University of the Third Age and Active Ageing*, International Perspectives on Aging 23, https://doi.org/10.1007/978-3-030-21515-6_12

Public meetings were held under the joint auspices of the Council for Adult Education and the Australian Council on the Ageing. The following guidelines established in the first few meetings have formed the blueprint for the organisational structure of U3A's in Australia: U3As would need to develop along self-help lines, involving all participants in every aspect of the development of their own learning; [ii] co-operation rather than competition would be encouraged between campuses; [iii] campuses should initially develop under the auspices of some established body as government funding for the education of older people was limited; [iv] there were to be no upper or lower age limits. Non-working older people would be advantaged by classes held during the day and a low yearly fee…and [v], the Australian Council on the Ageing was to provide an umbrella for the movement in Victoria and Australia as U3As grew. The first public meeting was held on 26 July 1984 at the City Centre of the Council of Adult Education, Melbourne. Despite limited advertising, 60 to 70 older citizens attended. The second public meeting, held in Hawthorn in November, was preceded by extensive media coverage. One of the reasons for choosing the Melbourne suburb was that Hawthorn housed more than double the national average of elderly residents. A quarter of its population was over 65.

Minichiello (1992: 412)

Australian U3As were founded as community-based, self-governed, not-for-profit groups and aim to provide affordable learning opportunities for third agers using the skills, experience, knowledge and abilities of the members themselves. Hence, in line with the Laslett's (1991: 178) principle that 'those who teach shall also learn—and those who learn shall also teach'.

In 2018, there were over 300 U3A groups in Australia, with a combined total of about 100,000 members (it is noteworthy that, collectively speaking, this is equivalent to 10% of all students enrolled in official universities in Australia), and these numbers show no signs of abating. U3As are generally found in the capital cities, as well as in regional and rural areas. Most are located on the eastern coastal fringe in towns that tend to be favoured as retirement centres. They may be large or small. For example, Sydney U3A has 6500 members, operating in seven metropolitan regional districts, and the Canberra U3A has 5,500 members. At the other end of the spectrum, there are small U3As of less than 50 members in small rural towns, where alternative opportunities for mental stimulation are rarely available. Since about 40% of Australians over the age of 65 live in regional and rural areas, it is not surprising that the majority of U3A members are found in such regions.

U3As in Australia are constituted as not-for-profit community associations, run by democratically elected committees. Administrative and ancillary duties such as webmasters, newsletter editors, course co-ordinators, as well as course leaders and tutors are undertaken by volunteers. Within that structure, U3As qualify as autonomous learning centres. As Beckett and Jones stated,

U3As in Australia are formed by groups of individuals in local communities, not organisations like many collaborative network organisations. They want a stimulating, participative recreational learning network for seniors, not a senior's social network that may provide learning. Both conventional and on-line learning modes are available. The focus is on long-term collaboration. Whilst each group operates as a separate entity, most choose to contribute to a regional virtual or physical office to cooperate in acquiring services as a group, sharing information on courses, interacting with governments and facilitating membership growth.

Beckett and Jones (2011: 195)

Most U3As offer a programme of weekly classes conducted in daylight hours and arranged on a school term or semester basis (many members have grand-parenting duties and so must be available for that during the school holidays). Some U3As operate every day of the week, others at lesser intervals as it suits the membership. U3As are responsible for their own expenses. They are funded primarily by annual subscriptions from members, on average about A\$50 per year, sometimes supplemented by top-up contributions for courses which may require special equipment, such as art classes, or by local 'fundraising' activities, usually social in nature. The subscription enables the member to attend any class offered by that U3A. In this way, U3A is affordable for those on low incomes. Some U3As are able to obtain one-off community grants for specific purposes, such as the purchase of computer equipment or sporting equipment, or the production of promotional material. U3As in regional areas might also receive in-kind support from their local Member of Parliament, for example, to print newsletters. In essence, however, U3As function through the volunteer efforts of their members, which at the turn of the twenty-first century Swindell, Vassella, Morgan, and Sayer (2011) described it two decades ago as a 4 million Australian Dollar bonanza:

> In total, 164 of 265 independent U3As in Australasia provided detailed counts of all activities carried out by their volunteers. In 2008, Australian U3As were supported by 871 000 volunteer hours and New Zealand U3As were supported by 69,000 volunteer hours. A notional value of \$20 per volunteer hour is rationalised. It values U3A volunteerism at \$21 million for Australia and \$1.9 million for New Zealand.
>
> Swindell, Vassella, et al. (2011: 196)

It is undoubtedly worth more since then. Most classes take place in rented accommodation, usually community centres and halls, libraries, or church buildings not utilised on weekdays. Some larger groups have been able to establish permanent office space, but smaller groups are usually administered from committee members' own homes. Some larger U3As in regional towns have been able to work with government authorities to obtain permanent space in buildings no longer required for their original purpose, such as demountable school buildings or railway stations. There is no consistent response by local government to support U3A. Some local government authorities with age-friendly policies are supportive and offer rental discounts or other incentives, others are less helpful.

Curricula of U3As

The curriculum of each U3A group is determined by the needs and interests of its members, the resources available to it and the availability of suitable tutors and small group leaders, all of whom are volunteer members of U3A.[1] Curriculum also varies widely in content and delivery, from lecture series to small group classes. Depending

---

[1] This is a major task in metropolitan areas such as Sydney and Canberra where the single U3A conducts several hundred courses every year. Some regional U3As conduct over 100 courses annually.

upon location, many U3A members provide lectures and courses for neighbouring U3As as well as their own. The advent of computer technology and the Internet has also enabled tutors and presenters to broaden their own expertise by researching new information to prepare courses on 'powerpoint' or via DVDs, as well as for U3As to access a wide range of downloadable courses. Within this framework, groups can offer a range of subjects, including art appreciation, astronomy, aboriginal culture, literature, history, music appreciation, film appreciation, theatre and drama, languages, philosophy, economics, law and science. Skills offered include drawing and painting, crafts, cooking, music and dance, yoga, tai chi, creative writing, photography, quilting and embroidery, and computer technology.

There is considerable research that demonstrates that music is an important contributor to well-being and self-esteem in older people, as a listener or a participant, especially with others (Peachey, Au, Caltabiano, Daniel, & Caltabiano, 2013; MacRitchie, 2016). In many U3As, retired music teachers and instrumentalists work hard to produce accomplished choirs and orchestras who compete in competitions or present at local community concerts. Many artists and photographers enter their work in competitions at all levels, confident in the skills acquired in U3A. Learning programmes on current affairs, philosophy and other discussion groups are also popular. Although the U3A is non-political and non-sectarian, members do not hold back on analysing and discussing major issues such as Animal Ethics, Euthanasia, Climate Change, the Future of the European Union, the Future of the United Nations, Trump's America, the Future of the Murray-Darling Basin,[2] the potential for an Australian Republic, or Reconciliation with Indigenous Australians—U3A encourages expression of a diversity of viewpoints. Some U3As have also challenged local secondary schools to debate (not overtly contentious issues), and to relate to other generational viewpoints. Light exercise, tai chi, yoga, bushwalking and cycling are other key class programmes. Many U3As also organise educational field trips for members, to venues such as art galleries and historical sites and museums. The range of interests of third agers, as well as their talents, enables them to take the roles of tutors or researchers, propelled by the ethos of 'learning for its own sake'. As one U3A member underlined, 'It's the ideal school—run by the students, you only study what interests you, and there are no exams!' (Ron, as cited in *The Senior*, 2017: 38).

Who Are the Members and Why Do They Join U3A?

U3A members come from different educational backgrounds and life experiences, although one notes that a relative majority of members hold a clear middle-class background. A survey undertaken by Hebestreit selected two U3As as a sample representative of the broader target population of all 64 U3As then in the State of Victoria and found that, of 627 respondents:

---

[2] The Murray-Darling Basin is Australia's major water catchment area in what is a very dry continent.

With regard to the level of education, 29.7% of the respondents had completed high school at the highest level of education, 23.4% had completed an undergraduate university degree, 29.8% had completed business/technical or trade certificates or undergraduate diplomas, and 11.2% held postgraduate degrees. With regard to occupation prior to retirement, 45.3% indicated that they were in professional occupations, 25% indicated secretarial/office work, and 14% indicated management positions.

Hebestreit (2008: 552)

Many U3A members have acquired tertiary educational qualifications for their past professional employment—teachers, librarians and health service professionals in particular—but there are no qualifications needed to join U3A. The range of locations across Australia from State capital and regional cities, and regional and rural townships, where U3As have been established, provides a wide range of 'second age' employment. There are many members whose previous employment has been in hospitality, service industries, home management, farming, trades or sales, whose zest for lifelong learning ensures their participation as valuable and valued members. Scattered among their numbers are some former university academics, civil servants and musicians, whose specialised experience and skills provide variety and expertise in the scope of courses and lectures available to their U3A. It is also noteworthy that, although post-World War II Australia has become a culturally diverse nation—for example the 2016 census revealed that more than one-fifth (21%) of Australians spoke a language other than English at home (Australian Bureau of Statistics 2017)—U3A membership is primarily of British heritage. A singular exception is the Melbourne Chinese Christian U3A.

Hebestreit's survey (2008) found 36% of respondents were aged between 60 and 69 years, 44.3% were aged 70–79 years and 14% were over 80 years. A survey of U3As in New South Wales undertaken in 2013 indicated that the ages of participants broadly range between 50 and 90, with 78% of members aged between 60 and 80 (de Hosson, 2013). Such patterns are similar in other States and Territories. In the wider population, at the lower end of the 60 s age range, many are still in employment, while those who retire in their early sixties—or earlier—are achieving other ambitions such as travel and are then coming into U3A for personal fulfilment. Hence, the largest cohort of membership tends to be in the age range of 70–80. Many U3As adopt a minimum age for eligibility for membership, usually 50 so as to equate to populist imagination of what constitutes 'ageing' with retirement from the labour market. Others take a broader view, given that Laslett (1991) himself took the view that the possibility has to be contemplated that the Third Age can be lived contemporaneously with the Second Age. While most U3A members will be approaching or going beyond the statutory retirement age, some U3As accept members who may still be in part-time employment, or younger persons who have had early retirement forced upon them through redundancy or who cannot work because of disability, but who also have a need for intellectual stimulation and social inclusion.

On average, about 70% of U3A members are women. To some extent, this may be because women have a greater life expectancy than men and are therefore more likely to be among the older age groups in U3A, while in Australian society men have traditionally been more likely to favour the company of other men and to belong to sporting and male-dominated social clubs. Moreover, the growing popularity of Men's Sheds, with their emphasis on mutual support and wood and metal craft and construction activities, is possibly more attractive to some men than U3A (Ahl, Hedegaard, & Golding, 2017). In response, some U3As have established their own Men's Sheds although in the interests of equality of opportunity, women are allowed both membership and entry in the specific learning programmes.

Undoubtedly, there are many clubs and organisations which can provide entertainment for retired people, but the U3A stands out as a main provider of opportunities for activities which promote lifelong learning, and which benefit the cognitive skills and general health of its members, and support positive ageing (Bridgstock, 2018). Research over several decades indicates the importance of maintaining the physical and mental health of older people in the community, in order to foster self-reliance, independence and well-being and to stave off mental deterioration (Cornwell & Waite, 2009). As well as their role of fostering lifelong learning for keeping seniors mentally active in a relaxed and informal manner, U3As provide the opportunity to develop those features within a friendly and supportive, social environment, leading to personal, mental, social and physical enhancement. As Swindell and colleagues' survey of U3As in Australia and New Zealand concluded, the U3A movement

> … quietly provided many, very-low-cost opportunities for members to take part in most or all of the successful ageing activities that are associated with continued independence in later life…Few organisations for retirees can point to a similar range of mentally, physically and socially stimulating courses, and the wide variety of volunteering opportunities provided by most U3As.
>
> Swindell, Vassella, et al. (2011: 200)

Hebestreit's survey found that the members surveyed regarded U3A as an important aspect of their lives leading to personal, mental, social and physical enhancement through the provision of lifelong education (Hebestreit, 2008). The same findings were echoed in de Hosson's (2013) survey in New South Wales which concluded that the principal reasons why people join U3A are wanting to learn new things, to participate in creative activities and to keep their minds active. Many claimed to have feelings of boredom following their retirement from employment, and U3A was perceived to provide, in the words of one U3A member, 'a reason to get up in the morning' (Paul, as cited in U3A Network NSW, n.d.). Equally important was the social interaction and the opportunity to develop friendships and companionship with people of a like mind. Another positive feature was affordability and 'value for money'.

Anecdotally, there are countless stories of new beginnings, especially though creative activities such as writing, painting, photography and music-making, which members may not have had time for during their working lives. Additionally, many U3A members achieve self-fulfilment by volunteering their time and talents to teach others, spending considerable time preparing lectures or classes for their U3A. An excellent example is a woman, a qualified exercise trainer, who contributed her professional experience to make a DVD of a light exercise 'balance and bones' strengthening programme which has enabled U3As and individual members to participate in a regular fitness series. This DVD has been especially valuable in rural and remote areas where access to such a professional programme is not often available.

As many persons in later life tend to be widowed or single, the U3A offers friendship and social inclusion as well as mental stimulation (de Hosson, 2013). Indeed, there is considerable evidence that social isolation can be detrimental to the physical and mental health of older people (Maginess, 2017), but the corollary is that positive social interaction enhances well-being in later life (Cornwell & Waite, 2009; Bridgstock, 2018). Moreover, non-formal continuing learning for older people can be liberating, by encouraging a sense of belonging to the wider world, and supporting them to maintain their independence, healthy dispositions, social interaction and intellectual functioning, in a convivial social environment to complement the learning environment.

U3A Online

U3A Online is the world's first 'virtual' U3A. It was established in 1998—with Australian Government funding during the International Year of the Older Persons with the intention of providing educational courses for older people, who may be isolated due to location or disability, and unable to join a 'terrestrial' U3A (U3A Online, n.d.). It has currently over 1000 members both within and outside Australia, and its courses are also utilised by U3As which are located in more remote parts of the country where access to such traditional learning programmes is limited. Like any other U3A, all the administration and teaching within U3A Online are carried out by retired volunteers:

> No one is paid. Governance and administration takes place through regular online meetings. Discussion and voting takes place by electronic forum, email and Skype. The annual general meetings, at which office bearers are elected, are also held online. In fact, it is not uncommon for volunteers to work closely together for many years, but never physically meet. All courses are written and taught by retired experts with the assistance of volunteer online editors…Each course runs for eight or nine weeks and is offered a few times a year, when the volunteer course leader is available. Participants interact with the leader and with others in the course by forum.
>
> Swindell, Grimbeek, and Heffernan (2011: 128)

At any time, U3A Online may have up to 50 learning programmes available, with facilitators interacting with learners, as well as including learners engaged in independent study and utilising 'chat rooms' for those who wish to meet online with others. U3A Online also maintains a Directory of U3As in Australia and New Zealand.

Statewide Networking

Australia is a federation comprising six States and two territories. Although U3As are autonomous bodies, in all States—except Tasmania—they have formed themselves into U3A Networks—the volunteer management committees of which are drawn from the U3A membership and which maintain low key administrations. The Australian Capital Territory has a single U3A for its 5,500 members. U3As in Tasmania, the smallest State, maintain informal networking between themselves and with interstate U3As. The Networks are not intended as hierarchical bodies, but function to provide mutual support for their member U3As. For example, public liability insurance and education copyright licences have been taken out by some State Networks as blanket cover for all of their member U3As, as it is cheaper for U3A groups to contribute to them than to take out individual policies and licences.

The Networks also provide advice and support on promotion and publicity, governance and relevant legislation. The Networks each run an annual or biennial conference where members can meet and share ideas and information, as well as provide an opportunity to hear keynote speakers on a range of important topics relevant to lifelong learning, positive ageing and research. In New South Wales, the Network also offers a Resource Library, accessible to U3A members anywhere in Australia via the Network website. The Library is made up of courses prepared and donated by U3A members, as well as providing access and links to other suitable courses. Although Laslett (1991) advised against accepting government grants, some State governments have recognised the value of U3A in assisting older people to maintain mental and physical health as part of preventative health strategies. In Victoria, for example, significant funding has been made available for several years and is applied to maintaining a central office and expanding U3A throughout the State. In other States, occasional small grants for specific purposes have been received, which have been applied to generic publicity for U3A including television advertising and supporting the establishment of new and small U3As. The Networks have formed U3A Alliance Australia (U3AAA) as the national face of the U3A movement in Australia. Due to the 'tyranny of distance', its deliberations are primarily conducted through electronic meetings, for the purposes of sharing information, and making submissions on national issues where relevant, although efforts are made to meet physically together at least once a year usually in conjunction with a State Network conference.

Community Engagement

Over time, U3A has engaged with the wider community. U3As have developed relationships with local government councils, local newspapers, radio and television stations and other older age group organisations in their localities. The State Networks are also building relationships with other State and national organisations with similar or complementary agendas. Several U3A groups have taken some of their activities into retirement villages for the benefit of those residents in care units. It is a misconception to think that if a person is frail and perhaps even bedridden that they lack interest in learning new things. The aim is to provide such people with the opportunity to participate in new learning, to enable them to engage in discussions, and to assist them to record their memoirs which can become a valuable history resource for family and for future researchers. U3A choirs often visit such units to provide musical entertainment where residents can join in community singalongs. There is considerable research which indicates that access to participation in music, dance and art is of particular value for older people, especially if suffering from dementia, as these activities stimulate feelings of happiness and well-being and open up new channels of communication (Cohen, 2000; Beaumont, 2013).

Although the U3A movement in Australia adopts the 'Cambridge' model, several U3As have also formed relationships with universities, which themselves are committed to policies of community engagement. These relationships can vary from the provision of access to university lectures on an audit basis (i.e. not as an enrolled student), to collaboration in university research projects. Academics readily accept invitations to make presentations to U3A groups about aspects of their research in a variety of disciplines. U3A members are encouraged to participate in university research programmes as willing participants in projects looking at various facets of gerontological research—such as anxiety and changes to brain activity in older patients, pharmacological treatment of ageing, and ageing and neuroplasticity. Many do so through online surveys, interviews, and in some instances, as simulated 'patients' for medical students.

Challenges for the Future

The primary challenge is: Will the U3A movement in Australia continue to grow? There is no reason to think it will not. As older members have passed on, new members will certainly continue to join as U3A's availability and response to its membership at grass-roots level ensures its relevance. Its volunteer base provides self-esteem to its tutors and committee members and affordability to its members. Moreover, the U3A movement in Australia has kept pace in the use of technology for both administrative efficiency and educational programme delivery. Most importantly, the fact that it offers intellectual stimulation and a range of activities within a social environment makes it attractive to retired people who want more than just bingo or crosswords or carpet bowls. Research on the impact of intellectual activities on brain health in later life suggests that lifelong learning is an important factor in 'preventative health' and in minimising the risk of dementia (Mestheneos & Withnall, 2016). Many older people wish to engage in meaningful activities and opportunities for new experiences, so one can only expect U3As to continue provide those opportunities for continued learning, not as an economic imperative but to enhance positive ageing.

Of course, there are practical considerations to consider. Although accommodation for classes and meetings can be difficult to locate, more so as membership grows and as rentals of suitable premises continue to rise, there is a strong belief among the members that individual U3As will adapt to their circumstances. Many voluntary bodies, especially those whose membership is of older people, find it hard to get members to take on committee roles and responsibilities, particularly when long-serving officers decide to retire. If U3A is to be maintained, and the demand is obviously there, it is important that those joining U3As understand the underlying U3A principles of mutuality and reciprocity and are prepared to contribute and to accept that all have a role to play, however small, in the effective operation of their group. Of equal, if not more, importance than its economic value is the value of volunteering to the volunteer and community spirit. Volunteering has been found to build collective efficacy by bestowing a sense of altruism and citizenship; developing political and negotiation skills; and inspiring people to work together to solve problems and take action to improve community life (Volunteering Australia, 2015). Additionally, research literature on volunteering indicates that unpaid community service is highly associated with positive levels of physical and psychological well-being in later life (Milbourn, Saraswati, & Buchanan, 2018).

Finally, while U3A is non-political, it is important to ensure that policy makers at all levels of government (Commonwealth, State and local) are aware of the U3A movement. Each year, Commonwealth government budget discussions tend to reinforce the idea that older Australians are a burden on the economy in their reliance on pensions and the greater part of government spending on health. The U3A movement should work to change such a perception and encourage policy makers to be cognitive of the value of U3A in maintaining mental and physical health for older people and in promoting lifelong learning and preventative health in general. It is equally important to remind government of the fact that the highest contribution of volunteer hours in Australia by seniors—aged between 65 and 84 years of age—delivers an economic value of millions of dollars to the community (Volunteering Australia, 2015). The U3A is an evolving and dynamic organisation and has the momentum to prosper further in the future. As Friedan (1994: 4) poignantly put it, 'ageing is not lost youth but a new stage for opportunity and strength'.

References

Ahl, H., Hedegaard, J., & Golding, B. (2017). How the Men's Shed Idea travels to Scandinavia. *Australian Journal of Adult Learning, 57*(3), 316–333.

Australian Bureau of Statistics. (2017). *Census reveals a fast changing, culturally diverse nation.* http://www.abs.gov.au/ausstats/abs%40.nsf/lookup/Media%20Release3. Accessed August 22, 2018.

Australian Bureau of Statistics. (2018). *2016 Census QuickStats: Australia.* http://quickstats.censusdata.abs.gov.au/census_services/getproduct/census/2016/quickstat/036. Accessed June 28, 2018.

Beaumont, G. (2013). *Arts activity proven to combat dementia.* https://www.limelightmagazine.com.au/features/arts-proven-to-combat-alzheimers/. Accessed July 13, 2018.

Beckett, R. C., & Jones, M. (2011). Active ageing: Using an ARCON framework to study U3A (University of the Third Age) in Australia. *International Federation for Information Processing, 362*, 189–196.

Bridgstock, M. (2018). *The impact of the Universities of the Third Age upon the health and welfare of their membership*. https://www.u3abrisbane.org.au/index.php/news. Accessed June 28, 2018.

Cohen, G. (2000). *The creative age: Awakening human potential in the second half of life*. New York: William Morris Paperbacks.

Cornwell, E., & Waite, L. (2009). Social disconnectedness, perceived isolation, and health among older adults. *Journal of Health and Social Behaviour, 50*(1), 31–48.

de Hosson, J. (2013). Profile of the network. In *Newslink*, U3A Network NSW Inc. (November 2013).

Friedan, B. (1994). How to live longer, better, wiser. *Parade Magazine*, March 20, 4–6.

Hebestreit, L. (2008). The Role of the University of the Third Age in meeting needs of adult learners in Victoria, Australia. *Australian Journal of Adult Learning, 48*(3), 547–565.

Laslett, P. (1991). *A fresh map of life*. Cambridge, MA: Harvard University Press.

MacRitchie, J. (2016). *Ageing in harmony: Why the third act of life should be musical*. https://theconversation.com/ageing-in-harmony-why-the-third-act-of-life-should-be-musical-57799. Accessed June 28, 2018.

Maginess, T. (2017). *Enhancing the wellbeing and wellness of older learners: A co-research paradigm*. Oxon, OX: Routledge.

Mestheneos, E., & Withnall, E. (2016). Ageing, learning and health: Making connections. *International Journal of Lifelong Education, 35*(5), 522–536.

Milbourn, B., Saraswati, J., & Buchanan, A. (2018). The relationship between time spent in volunteering activities and quality of life in adults over the age of 50 years: A systematic review. *British Journal of Occupational Therapy*, OnlineFirst. https://doi.org/10.1177/0308022618777219.

Minichiello, V. (1992). Meeting the educational needs of an aging population: The Australian Experience. *International Review of Education, 38*(4), 403–416.

Peachey, D., Au, A., Caltabiano, N., Daniel, R., & Caltabiano, M. (2013). *Music and ageing*, Inaugural Australian Psychological Society Conference, Cairns, 2013. https://researchonline.jcu.edu.au/30692. Accessed June 14, 2018.

Swindell, R., Grimbeek, P., & Heffernan, J. (2011a). U3A Online and successful aging: A smart way to help bridge the grey digital divide. In J. Soar, R. Swindell, & P. Tsang (Eds.), *Intelligent technologies for bridging the grey digital divide* (pp. 122–140). Hershey, New York: Information Science Reference.

Swindell, R., Vassella, K., Morgan, L., & Sayer, T. (2011b). Universities of The Third Age in Australia and New Zealand 2008: Capitalising on the cognitive resources of older volunteers. *Australasian Journal on Ageing, 30*(4), 196–201.

The Senior. (2017). *Never too many courses for this lifelong learner*, January 2017.

U3A Online. (n.d.). *U3A Online*. https://www.u3aonline.org.au. Accessed July 13, 2018.

U3A Network NSW Inc. (n.d.). *What U3A provides*. Promotional booklet.

Volunteering Australia. (2015). *Key facts about volunteering in Australia, 2015*. https://www.volunteeringaustralia.org/wp-content/uploads/VA-Key-statistics-about-Australian-volunteering-16-April-20151.pdf. Accessed July 3, 2018.

Ainslie Lamb AM, LLB, M.Ed., is a past President of the U3A Network New South Wales, Australia, and Chair of the U3A Alliance Australia. She practised as a lawyer for 25 years and in 1994 became foundation director and designer of the Practical Legal Training programme at the University of Wollongong. She retired in 2006 and joined U3A, where she has served in several committee roles, as well as making many presentations on a variety of legal, literary, art and reconciliation themes. She is an Honorary Professorial Fellow of the University of Wollongong and in 2003 was appointed a Member of the Order of Australia for services to community and legal education.

Chapter 13
The Development and Characteristics of Universities of the Third Age in Mainland China

Xinyi Zhao and Ernest Chui

> *Keep learning as long as you live.*
>
> Classical Chinese proverb

Introduction

Learning in later life has always been highly valued in China and is regarded as profoundly cultural in nature[1] (Field, 2012). However, older adult learning, without exceptions, reflects the social context where it takes place (Merriam & Mohamad, 2000). In China, a country where Confucianism has a profound impact on every aspect of life, citizens display a positive attitude towards education and learning, and there is high expectation that people, irrespective of their age, should pursue lifelong learning in order to become virtuous (Chui, 2012). Although in ancient times, most of the population was illiterate and excluded from participating in educational ventures, and only a limited number of people were able to take up learning in their later years, the idea of engaging in late-life learning has always been rooted in Chinese people's minds. In China, learning is indeed regarded as an act of virtue that benefits both the individual and society at large (Zhao & Chui, 2016). In Tam's words,

---

[1] This chapter introduces the development of Universities of the Third Age in the mainland areas of China and does not include the situation in Hong Kong Special Administrative Region, Macao Special Administrative Region, and Taiwan. Hence, throughout this chapter the term 'China' refers to the 'mainland areas' of China.

X. Zhao
School of Health Humanities, Peking University, Beijing, China
e-mail: zhaoxinyi@hsc.pku.edu.cn

E. Chui (✉)
Department of Social Work and Social Administration, Sau Po Center on Ageing, University of Hong Kong, Pokfulam, Hong Kong
e-mail: ernest@hku.hk

© Springer Nature Switzerland AG 2019
M. Formosa (ed.), *The University of the Third Age and Active Ageing*, International Perspectives on Aging 23, https://doi.org/10.1007/978-3-030-21515-6_13

In the context of lifelong learning, the Confucian philosophy conceptualizes learning as a lifelong process through which learners of any age learn to become a virtuous person and to lead a morally excellent life. Learning in Confucian terms is understood and practiced as learning for the sake of learning itself, as opposed to learning for the sake of an instrumental purpose.

Tam (2016: 193)

At the same time, the development of older adult learning in China was intertwined with socio-economic transitions (Zhao & Chui, 2016). Progressive economic development has enabled both central and local governments to devote more resources on promoting learning and welfare programmes for older adults. For older people themselves, disposable income has also improved significantly and pension reform has secured basic living standards (Chen, 2012). Improved economic status tends to encourage older adults to increase their expenditure on goods and services that enrich cultural life, including learning materials and programmes. Social development in China has influenced older adults to be more aware of how to lead healthier and more active lifestyles. As a result, participating in learning programmes has become a lifestyle choice for many older persons.

Historical Formations

In 1982, the Chinese government reformed the system of mandatory retirement for citizens working in government- and state-owned enterprises, which resulted in a large volume of older people experiencing retirement. This reform has been regarded as a key driver of older adult learning in China (China Association of Universities for the Third Age, 2010). Retirees who had a strong desire to be active after retirement pressed the government to establish and run learning programmes to participate in. In retrospect, it is clear that the government responded positively by viewing older adult learning as an optimal strategy to provide retirement activities for retirees. This resulted in the launching of learning institutions for older citizens, which could be regarded as the embryonic form of Universities for the Third Age (U3As) in China (Zhao & Chui, 2016). In China, the government takes the leading role in the provision of older adult learning by enacting laws, as well as by implementing policies and developmental schemas. Indeed, the right of older persons to access learning and education opportunities is guaranteed by law, with the *Law on Protection of Rights and Interests of Seniors* (1996) proclaiming that older adults have the right to continue their education, and that the country should develop older adult learning and also encourage society to operate different kinds of schools for older persons. In 2012, this law was revised so that the state would develop older adult education, as well as involve older citizens in the planning and coordination of the lifelong learning system.

In addition to legal statements, one also finds a range of policies on older adult learning set up by several departments of the central government but, most especially, the State Council, China National Committee on Ageing (CNCA), Ministry of Education, and Ministry of Culture. Such policies can be traced to the year 1999, as CNCA initiated a *Notice on CNCA Member Units' Responsibilities* (1999) which decreed that the Ministry of Culture will take responsibility for non-degree older adult learning across the country, as well as guide the ethos and U3A programmes. In the same year, the Ministry of Culture issued the *Recommendations on Strengthening the Work on the Older Persons' Cultural Life* (1999) to establish a network of older adult learning and establish U3As as an integral part of older persons' culture—thus making full use of the available range of social resources required to run older adult learning programmes. These recommendations also stipulated—for the first time—the need to explore the application of modern mass and social media for older adult learning. As a result, the goal of older adult learning was set to enhance knowledge, improve active and healthy ageing, and enable older people to take on volunteering roles (Chen, 2015). The Ministry for Culture's recommendations also embedded older adult learning under the auspices of public welfare, so that U3As began catering directly to the social needs of older Chinese citizens, without the need to charge excessive membership fees. Such policy directions are reminiscent of Groombridge's (1982) compelling reasons why education is highly relevant and crucial to an ageing population:

> Education promotes self-reliance and independence among the elderly. Education enables older people to cope more effectively in a complex and changing environment. Education for and by older people enhances their potential to contribute to society. Education encourages the elderly to communicate their experiences to each other and to other generations. Education is critical for lifelong learning and self-actualization.
>
> Groombridge (1982: 314)

In subsequent years, whilst the State Council's *Decision on Strengthening the Work of Ageing* (2000) stipulated that local governments should pay attention to the improvement of older adult learning and even develop distance education for older adults, the *Notice about Improving Elder Learning* (2001)—decreed by CNCA, Communist Party of China Central Committee (CPCCC), Ministry of Culture, Ministry of Education, and Ministry of Civil Affairs—required the Ministry of Culture to be fully responsible for the non-degree provision of older adult learning and for the promotion of older adult learning on the principle that all older adults are entitled to be given an opportunity to learn and contribute to society. The *Recommendations on Improving Grassroots Governments' Work on Ageing Issues* (2006), issued by CNCA, pronounced that all older adults are entitled to learning opportunities and set to guide older citizens towards a healthier lifestyle which is also characterised by resilient levels of active citizenship. It is also noteworthy that during this period, the quality of older adult education was improved by the distribution of better learning materials, incorporation of value-added teaching methods, and establishment of distance learning for older persons through the Internet, television, and radio programmes. The State Council's *White Paper on China's Ageing Issues* (2006) had the primary objective of improving the spiritual and cultural life of older persons—and hence, developing and advancing further the remit of older adult education. Such a policy approach is evocative of Moody's (1988: 194) transformative approach for late-life learning, one 'that enables people to grasp the meaning of their lives through history, philosophy, religion, or literature'. Instead of perceiving age as a deficit quality, Moody (ibid.) supported a developmental perspective which would emphasise gains as well as losses with ageing (Findsen & Formosa, 2011). In such an approach, the learning experience brings older persons' rich life experience to the classroom so as to facilitate the conversion of 'ageing' from an obstacle into a source of strength. Only so, Moody (1988) stated, will learning be successful in leading older learners to rediscover their identity as cultural bearers and culture creators.

The 2010s witnessed a steady publication of official policy guidelines targeting the expansion of lifelong learning in later life. The *Recommendations on Further Strengthening the Development of Older Adults' Culture* (2012)—issued by CNCA, the CPCCC, Ministry of Education, Ministry of Civil Affairs, Ministry of Finance, amongst other state institutions—decreed that lifelong education is a human right of older persons and that the requirement of realising lifelong learning opportunities for all, irrespective of age, qualifies as an important symbol of social progress. The recommendations also declared that the culture and education state departments should incorporate older adult learning into the lifelong and communal education system, so as to make full use of resources in communities and formal education institutions and provide educational activities according to older adults' needs and preferences and that U3As should renew themselves by offering courses in science, art, health care, and even practical skills to improve older persons' social adaptability and quality of life. A year later, the *Recommendations on Further Strengthening Preferential Treatment on Older Adults* (2013)—issued by no less than 24 state departments—pushed for equal treatment in culture, sports, and leisure facilities for older persons and stipulated that resources for older adult learning should be equally available to elders living in both urban and rural areas. It was also specified that public educational resources assist older adult learning, and most importantly, that tuition fees should be waived for 'poor' older learners in both non-formal and tertiary educational avenues. Finally, the State Council's *Recommendations on the Formulation and implementation of the Elderly Care Service Project* (2017) ruled that resources for older adult education should be made available to older adults in an equitable and orderly manner; it also recommended that tuition fees of U3As for 'poor' older persons be reduced further; encouraged urban and rural communities to locate suitable learning habitations for older adults; and called for further provision of suitable learning resources for older adults.

The government in China also issued a number of national plans that make special reference to older adult education. For instance, the *10th Five-Year Plan on China's Ageing Development* (2001) sought to develop older adult learning by designing and issuing national regulations which strengthen the carrying out of educational activities that are suitable for older adults, so as to enable them to increase their knowledge whilst also promoting active and healthy lifestyles. Subsequently, the *11th Five-Year Plan on China's Ageing Development* (2006) strove to develop older adult learning by forecasting the foundation of 10,000 new U3As by the year 2010, and thus, greatly improving the older adult learning network. It decreed that the national government, as well as all local authorities, should increase their investment in older adult learning and mobilise social forces to provide different forms of educational programmes, such as distance education and community education, especially for older adults living in rural areas as this has been found to mitigate against poverty. Whilst the *11th Five-Year Plan on Education Development* (2007) endeavoured to take full advantage of primary and secondary schools in the development of life-long learning and to expand the coverage of U3As, the *National Medium-and-long-term Plan for Education Reform and Development* (2010) emphasised the need to improve and strengthen the potential of community resources for older adult learning. The *12th Five-Year Plan on China's Ageing Development* (2011) strengthened older adult learning by exploring new models for learning, enriching the available curricula, increasing financial resources, and expanding the size of U3As. At the same time, this plan encouraged ideological education for elder learning by involving the Party branches, grass-roots governments, and age-friendly non-governmental organisations.

The main objective of the *Older Adult Learning Development Plan 2016* (*2016–2020*) was to establish—by the year 2020—a system of older adult learning with strong and wide network coverage, flexibility, and regulation, which would enable as much as 20% of older persons to participate in various forms of learning activities. Other targets included that every city should include a U3A; that 50% of townships should include opportunities for older adult learning; and that every province should develop a pool of resource persons that can sustain the development and carrying out of distance learning for older people. Moreover, it decreed that efforts should be made to enable cadres, professionals, and technicians to serve as part-time facilitators to achieve the target of one or two volunteer instructor teams for each U3A and to increase international collaboration, so as to learn from, and adopt, relevant overseas experiences. The latter would also serve to promote China's progress in older adult learning in foreign countries. Finally, the *13th Five-year Plan on China's Ageing Policy and the Older Adults Welfare System* (2017) stipulated that in the 2017–2022 period, one should ensure that the 50% target rate of townships having set up a U3A is achieved and that the proportion of older people who regularly participate in educational activities should reach 20% or more. The *13th Five-year Plan* also decreed the strengthening of the support to older adult learning by advocating that U3As operated by enterprises for their past retired employees should also open to the general public, improving the facilities of U3As, and ensuring a better seamless integration and interfacing between U3As on the one hand, and health, personal care, sport lifestyles, and cultural activities on the other. The Plan also supported pilot projects that integrate learning and care for older people and the publishing of learning resources for older adult education.

Current Circumstances, Curriculum, and Student Profile of U3As

U3As in China operate learning programmes that are not degree-based and operate under the guidance and support of relevant authorities such as the local old cadre bureau, education bureau, and/or cultural bureau (Yuan, 2011). U3A structures vary from place to place and are termed by local authorities as either 'universities' or 'schools':

> The difference between "university" and "school" mainly exists in scale: generally, those titled as "universities" are larger and are set up by the central, provincial, municipal or county governments or large corporations while the "schools" are usually in rural towns or urban communities on a smaller scale.
>
> Zhao and Chui (2016: 102–3)

Although U3As were originally established as schools for the older cadres, they are now open to the general public. For many years, many U3As were operated by state-owned institutions and enterprises; however, nowadays one finds many non-profit organisations designing a range of programmes in the realm of older adult learning. The majority of U3As are nonprofit, financed by the government, state-owned sectors, and donations (Xiao, 2000). Hence, older learners pay a very low tuition fee (ranging from tens to several hundred CNY) for each semester of learning programmes. Presently, the U3A movement has spread all over China with an exponentially increasing number of schools and enrolments. At the same time, U3As are no longer exclusively for old cadres but instead operate on a first-come-first-served basis. The first U3A in mainland China, Shandong Red Cross University for the Third Age, was established in Ji'nan City, Shandong Province, in 1983 (China Association of Universities for the Third Age, 2010). By 2013, the thirtieth year of U3A development, the number of U3As had grown to 43,000 and the number of learners had reached nearly 5 million (Gu, 2013). Till the end of 2017, there were around 490,000 elder schools with 7 million registered students (Ministry of Civil Affairs, 2018). The available data pertaining to numbers of schools and students from 1983 to 2017 is shown in Table 13.1. In terms of gender, more women than men participate in U3As—according to the China Association of Universities for the Third Age (2010) female participants accounted for about 65% of the total.

There is no doubt that China has the largest number of U3As in the world, with as much as 18 U3As including a student body of more than 10,000 members. In 2016, the three largest U3As in China were Xi'an U3A (200 study programmes, 60,000 members), Tianjin U3A (701 study programmes, 25,891 members), and Harbin U3A (566 study programmes, 20,299 members) (Teaching Study Department Guangzhou Elderly University, 2017). Reflecting on international literature on U3As, the formal education attainment of Chinese members is higher than the national average, and as most of the U3As are in urban areas older persons living in the countryside have less accessibility to participate in U3As. As the authors explained elsewhere,

> Many of the current cohort of older people had relatively low levels of education on average, probably due to the fact that some of them grew up during the wars (such as China's Resistance

Table 13.1 Number of U3As in China, by number of students (1983–2017)

Time	No. of U3As	No. of students
September 1983	1	586
December 1985	61	40,000
December 1988	916	124,800
October 1990	2091	218,400
March 1993	5331	471,000
July 1996	8378	695,900
October 1998	13,265	1,011,000
December 1999	16,676	1,413,000
September 2002	19,309	1,810,000
March 2004	25,060	2,313,600
April 2007	32,697	3,335,039
December 2008	40,161	4,302,365
December 2013	About 43,000	About 5 million
December 2017	About 49,000	About 7 million

Source China Association of Universities for the Third Age (2010); Gu (2013); Ministry of Civil Affairs (2018)

War Against Japanese Aggression from 1937 to 1945 and the Liberation War from 1945 to 1949) or in the period of the Cultural Revolution (1966–1976) when the opportunity to attend school was quite slim. The average years of educational attainment of older adults aged 60 and above are 5.9 for males and 5.8 for females. About 51.4% of the rural elders have no schooling and only 0.2% of them have a college-and-above level of education, while the percentages for urban older adults are 16.4% and 7.3% respectively.

Zhao and Chui (2016: 99–100)

With regard to the curriculum, the Chinese government encourages having 'two types of riches', referring to rich varieties of curricula and teaching strategies (Teaching Study Department of Guangzhou Elderly University, 2017). It is noteworthy that whilst in 1983, the first U3A provided only nine courses, recent literature reported that the number of study programmes had expanded from 200 in the 2000s to 350–400 (Lu, 2008). Some municipal U3As offer 168 (Harbin), 150 (Shanghai), and 127 (Wuhan) study programmes, respectively (Teaching Study Department of Guangzhou Elderly University, 2017). Whilst in the early days, the learning programmes were offered for leisure or recreation purposes, in due course the curriculum diversified to include life enrichment and promotion of good health and community service (Thompson, 2002). The learning programmes can be divided into the following categories: (1) current policies and laws; (2) literature and history: including classics literary appreciation, poetry writing, history, tourism, etc.; (3) language: including Mandarin, English, Japanese, etc.; (4) calligraphy and painting: including Chinese calligraphy, Chinese painting, oil painting, etc.; (5) arts and sports: including singing, choral, piano, Erhu, local drama, dance, tai-chi, chess, etc.; (6) health care: including Chinese medicine health care, nutrition, psychology, etc.; (7) family financial management: finance, securities investment; (8) household craft: including planting flowers, cooking, knitting, etc.; and (9) modern technology: including computers, photography, and image processing amongst other interests. Many U3As also classify courses into sub-classes ranging from primary level to advanced level. The learning programmes are usually held once a week, lasting from about 45 min to two hours each time.

Instructors, Teaching, and Learning Strategies

U3A study programmes are delivered by full-time or part-time lecturers, most of whom have a senior- or intermediate-level professional qualification recognised by the government (China Association of Universities for the Third Age, 2010). The lecturers and students create a mutual learning environment (Chen, 2017), and it is estimated that more than 60% of study programmes are taught through lectures, whilst others are taught through observation, practice, and/or discussion (Ding, 2017; Xu, 2015). The venues for U3As are provided by the government or state-owned enterprises/institutions. However, with the development of technology, some U3As began to provide long-distance education through television, radio-broadcasting, and the Internet, enabling older persons to learn from home. In 1995, the 'Shanghai Air U3A' started broadcasting on Shanghai Education Television Station. At present, there are about 260,000 older citizens in Shanghai who follow learning programmes broadcast by 'Shanghai Air U3A'. Beijing City, Zhejiang Province, and other places have also started to offer older adult learning programmes through distance education. Some other U3As also provide health-related courses through radio programmes. The recipients of radio education largely reside in towns and rural villages (China Association of Universities for the Third Age, 2010). In 1999, the 'Shanghai Online University for the Elders' was co-founded by Shanghai University for the Third Age, the Shanghai Municipal Committee on Ageing, and the Shanghai TV University and gave an opportunity to older persons to engage in older adult learning through the Internet. Subsequently, U3As in other cities also launched online universities, whereby they uploaded instructional videos and lecture texts to all older learners across the country.

Universities of the Third Age in China: Issues and Challenges

Despite rapid progress over the past several decades, U3As still face a range of issues and challenges, such as are found in other country and continental contexts (Formosa, 2012, 2014, 2016). First, the coverage of U3As needs to be distended. Here, it is necessary to mention that a large part of the future cohort of older adults will be the parents of an only child, as China's One Child Policy was initially implemented in 1979. Hence, if

> ...the only-children work in different regions from their parents, a greater number of empty-nested elder households will form and the time that older adults spend with the next generations will probably decrease. In this sense, engaging in learning activities would be a sound choice for them to enrich their later life. Therefore, elder education in the future should take such a social context into account, and planners provide courses that help seniors to adapt to role transitions so that they can enjoy themselves, even if the family structure becomes smaller.
>
> Zhao and Chui (2016: 107)

However, the majority of U3As are found in urban areas, limiting access to learning for older adults living in rural areas. Secondly, many learning programmes on the curriculum are related to the arts and are mainly meant for recreational purposes. As a result, there is a lack of learning programmes on vocational skills and areas unrelated to the arts and humanities. Thirdly, there is high turnover of U3A lecturers, and there are insufficient numbers of administrative staff so that U3As in China hold a staff-student ratio of 1:112 (Xu, 2015). Moreover, although central government bestowed the Ministry of Culture with almost complete authority over the administration of older adult learning in 2001, a range of administrative departments—including the education bureau, culture bureau, civil affairs bureau, and retired cadre bureau—all share management responsibility at a local level. This unclear and multi-sectorial administration system might easily lead to a certain degree of overlapping management responsibility, which could impact efficacy and efficiency (Zhou, 2004). Furthermore, to date, the central government has not established a unified management system for older adult education—and thus, there can be different management systems used in the development of late-life learning in different geographical regions. As concluded by the Teaching Study Department of Guangzhou Elderly University

> U3As in China possess distinct school-running characteristics and rich cultural deposits, but the connotative academic level is yet to be raised. In order to achieve sustainable development of U3As, it is required to further strengthen governmental support, increase education coverage, raise the level of theoretical research and internationalization, create an external environment that is looked upon seriously by society, and enhance modernization construction of U3As at all levels.
>
> Teaching Study Department of Guangzhou Elderly University (2017: 118)

Whilst miscellaneous approaches may provide more flexibility to different U3As ensuring that their specific management mechanisms cater for and are sensitive to their respective local contexts, such circumstances may result in inconsistencies and wide disparities between regions, municipalities, and localities.

Contribution of Chinese Universities of the Third Age to Active Ageing and Future Challenges

It is believed that older adult learning can noticeably inhibit cognitive decline and improve the cognitive abilities of old people (Boron, Turiano, Willis, & Schaie, 2007). Late-life learning also has the potential to counter mental tribulations and improve psychological well-being (Morrow-Howell, Kinnevy, & Mann, 1999). Hence, it is not surprising that scholars in China have found that U3A learning has a significant positive impact on older people's quality of life (Chen, 2010; Yang, 2011; Zhou & Zhang, 1998). U3As can improve old people's social engagement and strengthen their contributions to communities through voluntary activities which, in turn, promotes active ageing. In practice, some U3As frequently assist their members to join community services and voluntary activities outside the classroom, leading their members to experience higher levels of productive ageing. Older people's active participation in learning can also be seen as making an important contribution to society (Du & Wang, 2013). This is grounded on the Confucian emphasis on the virtue of learning which is regarded as essential for becoming a virtuous person (Kim & Merriam, 2004). Learning has its intrinsic value per se, and learners are not only highly respected but they are morally obligated to serve the community. In Chinese culture, everyone—including older citizens—is expected to pursue lifelong learning, as far as is practically feasible, to accomplish personal fulfilment (Chui, 2012).

In recent decades, population ageing and socio-economic development in mainland China have functioned to expand the development of U3As beyond any imaginable prospects, a state of affairs which shall certainly continue in the foreseeable future. According to the State Council's *Plan of Elder Education Development (2016–2020)*, issued in 2016, and *The 13th Five-year Plan on China's Ageing Policy and the Older Adults Welfare System,* issued in 2017, the government has set up a number of important goals. It is expected that by the early 2020s, there will be a participation rate of 20% amongst older people in various forms of educational activities. It is also envisaged that every city should have a U3A and 50% of townships (streets) should include some type of late-life learning initiative. This would mark the establishment of a system of elder learning with wider coverage, more diversified courses, better learning resources, and more qualified teaching staff. Looking forward, there may be the need for the Chinese government to devise a more systematic approach in promoting late-life learning via U3A and to learn from the international community, so as to benchmark with international standards and practices, and enable China's looming large older population to benefit more from lifelong learning.

References

Boron, J. B., Turiano, N. A., Willis, S. L., & Schaie, K. W. (2007). Effects of cognitive training on change in accuracy in inductive reasoning ability. *The Journals of Gerontology Series B: Psychological Sciences and Social Sciences, 62*(3), 179–186.

Chen, J. (2012). To emphasize and develop the function of social security in income redistribution. In Y. Wang (Ed.), *China social security system development report, No.5: 2012 social security and income redistribution* (pp. 1–6). Beijing: Social Sciences Academic Press (in Chinese).

Chen, L. K. (2015). Rethinking successful aging: Older female volunteers' perspectives in Taiwan. *Asian Journal of Women's Studies, 21*(3), 215–231.

Chen, M. (2010). The influence of music learning on the subjective wellbeing of the elderly. *Journal of Xinghai Conservatory of Music, 121*(4), 86–92.

Chen, M. (2017). An analysis on the status and development strategies of the Universities for the Third Age. *Inner Mongolia Education, 14*, 4–5.

China Association of Universities for the Third Age (Ed.). (2010). *Research on the elderly education in urban China*. Beijing: Higher Education Press (in Chinese).

Chui, E. (2012). Elderly learning in Chinese communities: China, Hong Kong, Taiwan and Singapore. In G. Boulton-Lewis & M. Tam (Eds.), *Active ageing, active learning: Issues and challenges* (pp. 141–161). Dordrecht, Netherlands: Springer.

Ding, Z. (2017). Current status, problems and strategies of the Universities for the Third Age. *Modern Distance Education, 4*, 70–74 (in Chinese).

Du, P., & Wang, F. (2013). Productive ageing in China: Development of concepts and policy practice. *Ageing International, 38*(1), 4–14.

Field, J. (2012). Lifelong learning, welfare and mental well-being into older age: Trends in policies in Europe. In G. Boulton-Lewis & M. Tam (Eds.), *Active ageing, active learning: Issues and challenges* (pp. 11–20). Dordrecht, Netherlands: Springer.

Findsen, B., & Formosa, M. (2011). *Lifelong learning in later life: A handbook on older adult learning*. Rotterdam, Netherlands: Sense Publishers.

Formosa, M. (2012). Education and older adults at the University of the Third Age. *Educational Gerontology, 38*(1), 114–125.

Formosa, M. (2014). Four decades of Universities of the Third Age: Past, present, and future. *Ageing & Society, 34*(1), 42–66.

Formosa, M. (2016). Malta. In B. Findsen & M. Formosa (Eds.), *International perspectives on older adult education: Research, policies, practices* (pp. 161–272). Cham, Switzerland: Springer.

Groombridge, P. (1982). Learning, education, and later life. *Adult Education, 54*(4), 314–325.

Gu, X. (2013). Opening address on the plenary meeting of International Association of Universities of Third Age. *Education Journal for Senior Citizens, 6*, 6–7.

Kim, A., & Merriam, S. B. (2004). Motivations for learning among older adults in a learning in retirement institute. *Educational Gerontology, 30*(6), 441–455.

Lu, J. (2008). The summary of the theoretical research on elder education in China in the past 25 years. *Education Journal for Senior Citizens, 12*, 15–22 (in Chinese).

Merriam, S. B., & Mohamad, M. (2000). How cultural values shape learning in older adulthood: The case of Malaysia. *Adult Education Quarterly, 51*(1), 45–63.

Ministry of Civil Affairs. (2018). *Statistical bulletin for social development 2017*. From http://www.mca.gov.cn/article//sj/tjgb/201808/20180800010446.shtml. Accessed September 01, 2018 (in Chinese).

Moody, H. R. (1988). *Abundance of life: Human development policies for an aging society*. New York: Colombia University Press.

Morrow-Howell, N., Kinnevy, S., & Mann, M. (1999). The perceived benefits of participation in volunteer and educational activities. *Journal of Gerontological Social Work, 32*, 65–80.

Tam, M. (2016). The Confucian view of lifelong learning: Relevancy to the teaching and learning of older adults. In C. M. Lam & J. Park (Eds.), *Sociological and philosophical perspectives on education in Asia-Pacific region* (pp. 193–204). Dordrecht, Netherlands: Springer.

Teaching Study Department Guangzhou Elderly University. (2017). On history and development of universities of the third age in China. *Roczniki Nauk Społecznych, 2*, 101–118. https://www.ceeol.com/search/article-detail?id=613566. Accessed September 2, 2018.

Thompson, J. (2002). *The amazing University of the Third Age in China today*. http://worldu3a.org/resources/u3a-china.htm. Accessed December 5, 2011.

Xiao, C. (2000). *China: Lifelong learning and the use of new technology*. http://www.techknowlogia. org/TKL_Articles/PDF/171.pdf. Accessed May 10, 2013.

Xu, Z. (2015). The analysis and countermeasures on the current situation of university for the elderly: Based on the empirical study of Hangzhou City. *Adult Education, 04*, 34–36 (in Chinese).

Yang, Q. (2011). Old adult education is an important cause of coping with population aging. In J. Sun (Ed.), *Education for the elderly in China: Research and practice* (pp. 33–40). Beijing: Science Press (in Chinese).

Yuan, X. (2011). Some issues about the development of older adult education. In J. Sun (Ed.), *Education for the elderly in China: Research and practice* (pp. 3–9). Beijing: Science Press (in Chinese).

Zhao, X., & Chui, E. (2016). Mainland China. In B. Findsen & M. Formosa (Eds.), *International perspectives on older adults education: Research, policies and practice* (pp. 99–109). New York: Springer.

Zhou, L. (2004). *Analysis on the elderly education in urban communities* (Unpublished Master degree dissertation). Minzu University of China, Beijing (in Chinese).

Zhou, S., & Zhang, L. (1998). Comparative analysis on the psychological and physical status of 201 students in senior university before and after the entrance. *Medicine and Society, 11*(5), 35–37 (in Chinese).

Xinyi Zhao Ph.D., is Assistant Professor at the School of Health Humanities, Peking University. She is also the Honorary Research Fellow of Sau Po Centre on Ageing, University of Hong Kong. Her research fields include productive ageing, lifelong learning, long-term care and medical insurance.

Ernest Chui Ph.D., EdD, RSW, is Associate Professor at the Department of Social Work and Social Administration, and former Director of the Sau Po Center on Ageing of the University of Hong Kong. He had served as Vice-Chairman of the Hong Kong Social Workers Registration Board. His teaching and research interests cover social work education, social policy, elderly welfare, housing and community work. His publications appear in academic journals and include *Social Work Education, International Journal of Social Welfare, Asia Pacific Journal of Social Work and Development, Housing Theory & Society, Australasian Journal on Ageing, Habitat International*, and *Community Development Journal*.

Chapter 14
Third Age Learning in Hong Kong: The Elder Academy Experience

Maureen Tam

Introduction

Hong Kong is an international city located on the coast of southern China. Because of its colonial ruling by the British for almost a century, Hong Kong is considered as the crossroads between East and West. The former British colony has since 1 July 1997 become the Hong Kong Special Administrative Region (HKSAR) of China following the handover of sovereignty from the British to the Communist Chinese Government. Hong Kong has a total population of 7.41 million in 2017 (Census and Statistics Department, 2017). In recent decades, the population has shown an increasing trend of ageing. The median age of the population rose from 28.8 in 1986 to 43.4 in 2016. On the contrary, the fertility rate decreased from 1367 births per 1000 women in 1986 to 1205 in 2016. Life expectancy for men has increased from 74.1 years in 1986 to 81.3 years in 2016 and from 79.4 years to 87.3 years for women during the same period. The combination of a low fertility rate and longer lifespan means that Hong Kong is having a rapidly ageing population, where the proportion of the population aged 65 and over rose markedly from 7.7% in 1986 to 15.9% in 2016. This cohort is expected to rise further to 31.1% in 2036 and to 36.6% by 2066 (ibid.). As a result, Hong Kong follows other middle- and high-income countries in having to face the challenge of a growing ageing population, requiring the government to put in place policies and plans to help its older citizens to age actively and happily. Of these policies and plans, the promulgation of lifelong learning is believed to play a vital role in promoting the quality of life in later years.

M. Tam (✉)
Education University of Hong Kong (EdUHK), Tai Po, Hong Kong
e-mail: msltam@eduhk.hk

© Springer Nature Switzerland AG 2019
M. Formosa (ed.), *The University of the Third Age and Active Ageing*, International Perspectives on Aging 23, https://doi.org/10.1007/978-3-030-21515-6_14

Third Age Learning in the World

The fast ageing of populations is a worldwide issue. The demographic imperative of the ageing population has an impact on many aspects of society. It affects government policy and plans on social welfare, housing, medical, health care, and retirement protection. More recently, there is another need which is receiving increasing government attention. This need is education in the form of third age or lifelong learning for older persons. In recent decades, increasing interest and attention were paid to the need for older adults to engage in learning. More and more so the benefits of learning for older adults are being recognised and highlighted by research. Over the past two decades, there has been rapid expansion of third age learning opportunities for older adults in many parts of the world. Such expansion is supported by government policies and resources, though limited in many cases. Third age learning takes a variety of forms across the globe. There are, in general, four types of provision that caters to third age learning (Findsen, 2002). The first type is those self-help agencies which are run and controlled by older learners themselves. The second type includes agencies that specifically provide learning programmes for older adults. The third type is made up of providers in mainstream education that offer some programmes relevant to older learners. The fourth type is no provision at all for older adults to engage in later life learning so that facilities and policies are totally lacking.

Over the past few decades, there has been a proliferation of lifelong learning opportunities catering exclusively to the interests of older adults. The growth is due to a few worldwide trends, notably population ageing, declining birth rates, longer lifespans, a stronger emphasis on a knowledge-based society and rising levels of education among older adults who expect to continue to engage in learning in a bid to stay active and healthy as they age. Previous research has shown that continued learning leads to many positive outcomes. Cognitively, learning helps to gain new knowledge, keep up the activity level, challenge the mind, thus keeping mentally and physically fit (e.g. Formosa, 2016; Narushima, 2008; Tam, Aird, Boulton-Lewis, & Buys, 2016; Withnall, 2010). Socially, learning activities provide a very important source of building or maintaining social relationships (e.g. Aberg, 2016; Brownie, 2014; Jamieson, 2007; Luppi, 2009; Schuller, Preston, Hammond, Brassett-Grundy, & Bynner, 2004). Psychologically, learning helps to boost confidence and self-esteem of older persons, who have a greater propensity to enjoy life and feeling fulfilled (e.g. Boulton-Lewis & Buys, 2015; Jenkins, 2011; Lun, 2011; Villar, Pinazo, Triado, Celdran, & Sole, 2010).

Lifelong learning for older adults is generally better known as 'Third Age Learning' in the field since the first University of the Third Age (U3A) was founded in Toulouse, France, in 1973 (Swindell & Thompson, 1995). Since then, third age learning has evolved and developed into various forms across the globe. To differentiate the third age from other stages of life, Laslett (1989) offers a four-phase model of the human lifespan. First is the stage where a person is dependent on others. Second is one of maturity where a person usually assumes the roles of husband/wife and parents and is working to build a career and maintain financial independence. Third is the age of autonomy and freedom to pursue one's interests and lifestyle that is rid of many responsibilities imposed in the second age. Fourth is the age of dependency imminent to death. Here, Laslett has defined the third age in a positive light, emphasising capacity rather than deficiency. It is a widely accepted definition in the field where third agers are portrayed as active, autonomous persons with opportunities to build on their years of knowledge and experience to enhance their individual capabilities (Talmage, Lacher, Pstross, Knopf, & Burkhart, 2015).

Due to its French origin, the first U3A founded in Toulouse is known as the French model (Swindell & Thompson, 1995). In this model, a U3A needs to be associated with a traditional university, which provides negotiated access to its regular university courses, other learning activities and even campus facilities. Some U3As receive funding from the university, some receive subsidy from the local government, whilst some others charge fees paid by U3A members themselves. A departure from the French model was the British model of U3As as the idea of third age learning later spread to England. The first U3A of the British model was established in connection with Cambridge University in 1981 (ibid.). The British model follows a self-help approach with classes and courses being organised by members of the U3A, which does not have to be associated with a traditional university. Each U3A is independent in terms of funding and management, which is run by its own elected management committee and is funded by charging minimal membership fees. Very often, classes are held in community halls, libraries and even private homes with flexible timetables, negotiated curricula and teaching modes with no assessment nor admission requirements. There is usually a wide course variety where teachers and students often swap roles as members can be both teachers and students, forming a learning community to share knowledge and experiences (Formosa, 2014, 2016).

Since the establishment of the first U3A in Europe, third age learning has received increasing attention and interest across the globe. Different versions of third age institutes have emerged and mushroomed. Some notable examples include the Elderhostel and Osher Lifelong Learning Institutes in the USA (Brady, Cardale, & Neidy, 2013); the University Programs for Older Adults in a few Spanish-speaking countries (Brownie, 2014); the Evergreen College and the Universities for Older Adults in Taiwan (Lin & Huang, 2013); and the Senior Centres Without Walls in Canada (Newall & Menec, 2015). Despite the different names and labels, U3As or U3A-like organisations in different parts of the world have grown exponentially over the past few decades since the establishment of the first U3A in France in the 1970s. There are no official statistics as to how many U3As currently exist in the world. However, in an earlier chapter in this volume Formosa provided a contemporary snapshot on how the numbers of U3As and members have burgeoned in some countries:

Australia included some 300 U3As, with a membership around the 100,000 mark, whilst its neighbour New Zealand held 84 U3As with the members of the 25 Auckland U3A community numbering 3,719 in 2017. Figures for Britain reached over 1,000 U3As (400,000 members) at the end of 2017, and a 2013 Interest Group Survey revealed that there are in excess of 36,000 U3A interest groups in the region…In the Asian continent, China alone included 60,867 U3As and around 7,643,100 members in 2015.

Formosa, 2019: XX

These figures are certainly revealing but are just the tip of the iceberg, which are expected to soar as the demand for third age learning is ever increasing due to the fast-ageing world populations.

Third Age Learning in Hong Kong

In Hong Kong, third age learning first emerged in the late 1980s where non-government organisations (NGOs) played an active role in initiating and organising lifelong learning programmes and activities for older persons at the local community level (Elder Academy, 2018). Similar mention was made by Leung, Lui, and Chi (2006) who also stated that learning for older adults in Hong Kong began in 1984. At that time, programmes were organised by older adult community centres and social services organisations where courses were mostly personal development in nature, requiring neither formal assessment nor admission qualifications (Zhang & Ha, 2001). Most of these non-formal learning opportunities are addressing the leisure and practical needs of older adults. Examples included computer courses, tai chi lessons, Chinese calligraphy and painting, and dancing and singing, among others. This period of older adult learning was thus characterised by a non-formal approach with little government involvement from the 1980s through the early years in the 1990s (Lee & Chan, 2002). This has remained unchanged until the establishment of the Elderly Commission of Hong Kong in 1997, whose mandate is to advise the new HKSAR government on policies and issues related to the growing ageing population in Hong Kong.

Lifelong learning, among a host of other pressing ageing issues such as ageing welfare, housing, medical and health care, falls under the auspices of the EC. The EC is aware of the benefits of lifelong learning to active and healthy ageing and has therefore advised the government since 1997 to adopt a more coordinated and active approach to the promotion of third age learning among older adults to help them enhance quality of life as they age. To this effect, Tam (2012) has delineated the development of third age learning in Hong Kong in two distinct stages. The first stage referred to the pre-1997 period before the of Hong Kong was set up and where third age learning was mainly non-formal in nature and offered by individual social service agencies at the local community level. The second stage referred to the post-1997 period where the founding of the of Hong Kong has changed the landscape of third age learning development in Hong Kong. Since then, lifelong learning for older persons is promoted and encouraged by the government through social campaigns, projects and limited funding support to providers. And the approach is a more coordinated one underpinned by policies and directives from the Elderly Commission. In its *Report on Healthy Ageing*, the Elderly Commission of Hong Kong (2001) identified 'lifelong learning by the elderly' as one of the four strategic imperatives for realising active and healthy ageing for older adults in Hong Kong.

The Elderly Commission of Hong Kong has been active in crafting policies and plans on numerous fronts since its establishment after 1997. With respect to third age learning, one most notable policy initiative by the Commission is the establishment of the Elder Academy Scheme jointly with the government's Labour and Welfare Bureau in early 2007. The aim of the Scheme is to coordinate efforts and provisions of lifelong learning for older adults in Hong Kong under the collective banner of the Elder Academy Scheme. By joining the Scheme, providers from any sector will receive seed funding to set up an Elder Academy forming a network of academies across all 18 districts in Hong Kong. As stated on the Elder Academy (2018) website, the initial plan was to set up elder academies mainly in primary and secondary schools. In due course, the network was expanded to include tertiary institutions with different providers catering to a variety of needs of older adults for continued learning at different levels. The network was initially made up of 78 elder academies established in primary and secondary schools. To date, the number has grown to 140, and the Scheme has expanded to include six universities in the tertiary sector. According to the government's response to the Legislative Council on 29 June 2016, it was reported that there is a total number of 10,000 learning places each year provided by all elder academies combined (Labour and Welfare Bureau of Hong Kong, 2016). Joining the network is voluntary, but it requires an application process. Once approved, each new Elder Academy will be provided with a small grant from the government as seed funding. Schools, primary and secondary, are encouraged to partner with local volunteer organisations or non-governmental organisations to offer programmes and courses that can make good use of the school campuses and the participation of school students for the promotion of intergenerational learning. To be members of the network, an Elder Academy is required to comply with the following objectives (Elder Academy, 2018):

1. To promote lifelong learning by encouraging older persons to make best use of their time and to keep pace with society through acquiring new knowledge and learning new skills.
2. To maintain healthy physical and mental well-being by enhancing their sense of achievement and self-confidence through learning.
3. To foster a sense of worthiness in later life by offering a platform for older persons to share knowledge, demonstrate creativity, serve the community and continue to make a contribution to society.
4. To optimise existing resources through partnerships with schools which are equipped with the requisite resources and facilities for learning that can take place after school hours and at weekends.
5. To promote integration between older persons and younger peers by engaging them in learning activities for intergenerational harmony and rapport.
6. To strengthen civic education by having uniformed groups such as the Scout Association to offer volunteer services to older persons to promote civic education and to foster community spirit.
7. To promote cross-sectoral harmony through collaboration among schools, tertiary institutions and non-governmental organisations which include the District Elderly Community Centres.

Two years later, in 2009 the Hong Kong government established the Elder Academy Development Foundation under the auspices of the Elderly Commission with an aim to further support lifelong learning for its older citizens. The Foundation provides funding for projects, programmes and activities that promote continued learning for older persons to keep pace with the times, to stay active physically and mentally and to contribute to society. Most of the courses run by elder academies at the community level target older adults aged 60 and above. They are informal in terms of both content and mode of study. Courses are mainly for personal interest and development, with participation being mostly on a part-time status and voluntary condition. Elder academies are also established in universities or tertiary institutions, though they are study programmes which are linked to some university classes. Opportunities are provided for older adults to enrol in university courses as auditing students, who do not take part in any assessment and neither will earn credits towards a degree qualification. As to the curricula, there is a wide range of offerings from elder academies at various levels to meet the different needs, interests and abilities of older persons. The aim is to help raise their quality of life and capability of adjusting better to ageing. To this end, a wide range of learning opportunities are made available, which include both academic learning and leisurely pursuits such as art and crafts, physical activities like dancing and sports. There are usually no admission requirements or examinations for assessment. The whole idea is to encourage active ageing through active learning, which should be made barrier-free and stress-free for older learners.

In addition to the Elder Academy Network, there is also the U3A Network in Hong Kong co-founded by a business company and a social service organisation in 2006. The aim of the U3A Network of Hong Kong is to encourage retirees to stay active through engagement in lifelong learning and volunteering. Different from the EA Network, the U3A Network is more akin to a 'self-initiating', 'self-learning', 'self-teaching' and 'self-managing' model of third age learning (Hong Kong Council of Social Service, 2014). Forming the network is a group of older service agencies which receive funding support to operate self-learning centres for older adults to learn and to organise classes for their peers. After a decade of development between 2006 and 2016, the network has established 48 self-learning centres, providing more than 4600 courses with over 70,000 learning opportunities for third age learners (ibid., 2016). Though smaller in scale and without government involvement, the U3A Network of Hong Kong is running in parallel with the Hong Kong Elder Academy Network to jointly promote third age learning among older adults, who are encouraged to pursue interests, fulfil their dreams and contribute to the community.

Characteristics of the Elder Academy Network

This model of third age learning in the form of a network of elder academies is claimed to be unique with Hong Kong characteristics (Elder Academy, 2018). It adopts a cross-sectoral, collaborative approach where the government plays a coordinating and supporting role to encourage providers from various sectors to offer, jointly or separately, learning opportunities that cater to the various learning needs of older adults. It is the cross-sectoral linkages and the coordination of interaction, connection and cooperation among and between the government and providers in such a network of educational provision that sets the Elder Academy Network apart from the other models of third age learning in the world (Tam, 2012). According to Brady et al. (2013), lifelong learning institutes can be categorised into two main types: institution-driven and member-driven. If it is institution-driven, the curriculum is planned by professional staff and taught by regular instructors employed by the institution. However, in the member-driven context, courses of study are planned and taught by the members themselves. The Elder Academy Network in Hong Kong follows neither of these models. The network, claiming to be distinct and unique with Hong Kong characteristics, is characterised by the co-investment and joint engagement between the government and the various stakeholders. Together they form a network comprising schools (primary and secondary), universities and non-government organisations, whilst also providing a myriad of offerings at different levels that cater to the wide-ranging interests and capabilities of older adults. The government plays a very important role in the network by providing clear and effective coordination of efforts, interaction and cooperation across elder academies in Hong Kong. It is certainly the cross-sectoral linkages and the network of a wide range of providers that have made the Elder Academy Network a unique model of third age learning.

Another distinct feature of the Elder Academy Scheme is its emphasis on intergenerational learning. With its origins first started in schools, the Scheme's main focus is to promote intergenerational interactions between older persons and school children, linking the older and younger generations through a series of purposive interactions with reciprocal benefits (So & Shek, 2011). To a large extent, intergenerational learning is also happening inside the university classrooms, where younger undergraduate students will learn together with older persons who also take part in class discussions with them, resulting in intergenerational sharing of views and experiences. Unlike many U3As or U3A-like institutes in the world which target mainly older persons in their curricula and provisions, elder academies in Hong Kong aim to achieve intergenerational learning and harmony to result in mutual benefits for the old and young. Intergenerational learning, according to Manheimer (1997), is a co-learning process where the old and young learn new skills and gain insights into the lives of themselves and their counterparts, promoting cross-generational understanding, solidarity and trust.

Criticisms of the Elder Academy Network

Though ideal it might seem, the Elder Academy Network is not without criticisms. Current funding for elder academies is irregular and non-recurrent, small and inadequate, which has compelled providers to compete with each other for the limited resources and to operate on a market-oriented basis. Providers very often find themselves facing a tension between a commitment to the provision of elder learning as a community service and the need to operate on a self-financing basis through competition in the marketplace. Because of the need of staying competitive or simply surviving in the market, providers have to choose between offering courses that will address the immediate wants of the elder learners and providing the kinds of knowledge and skills that are needed in the long run. To this effect, providers are facing the dilemma of providing learning for communitarian values, on the one hand, and offering learning opportunities that can be readily accepted by the market, on the other.

Cognisant of its fast-ageing population—due to declining birth rates and longer lifespans—the Hong Kong government has since 2007 developed and implemented policies and plans to facilitate older adults' engagement in third age learning, in particular, through a network of elder academies across different parts of Hong Kong. Eleven years on since its establishment in 2007, there remained a lack of information if the policy, provision and practice have been effective, adequate and valued from the perspectives of different stakeholders, including policy makers, providers and elder learners themselves. So far, there is limited information as to its effectiveness and impact on stakeholders. Research upon an evaluation of the Elder Academy Scheme is needed to evaluate how effective the initiative has been in encouraging older adults to engage in third age learning. Over the past eleven years, there has been a rapid expansion in terms of the number of elder academies in the network, an increased range in the types of programmes available, and ever-increasing participation rates among older persons in Hong Kong. This has occurred in the absence of any comprehensive evaluation or assessment. There is a need to address this crucial gap by critically examining the efficacy, adequacy and value of the policy, provision and practice of how third age learning is planned and organised in Hong Kong. It will provide invaluable feedback to the Elder Academy Scheme from stakeholders to determine if its objectives and targets are being met and will provide important information for further improvement and development.

Conclusion

Although the Elder Academy Scheme looks like an effective approach of supporting and providing third age learning in Hong Kong, there are challenges that might impede its future development. In Chui and Xhao's words,

> It is anticipated that in the coming decade the current soon-to-be-old or baby-boomer cohort would constitute the major market for lifelong learning activities. This future cohort of seniors should be relatively better off than the current cohort in terms of health, financial situation, literacy and personal efficacy. Their aspirations for learning may not be satisfied by the sheer provision of conventional, classroom, teacher-oriented, leisure or interest oriented types of learning activities currently provided by the majority of NGO providers.

> Chui and Zhao (2016: 17)

One imminent challenge is the lack of stable and sustainable funding and support from the government for providers and elder learners through the Elder Academy Scheme. Moreover, there is a need of policy to shift from the market-oriented provision to one that supports third age learning as a necessity for people, regardless of age, ability and background, to engage in simply because they have the natural capacity and desire to continue to learn throughout one's lifespan. Older persons possess also the right to learn and interact with others for the purpose of personal growth and social relationships. It is therefore paramount that the Elder Academy Scheme be critically reviewed for effectiveness to ensure it is meeting the needs and expectations of older people and supporting providers in delivering high-quality programmes to older persons. There is a lack of evaluative study or research to investigate the effectiveness of current policies and provisions of third age learning in Hong Kong. In conclusion, to determine whether such policies and provisions have achieved the desired outcomes, they must be subjected to rigorous evaluation. Only then can these policies and provisions claim to have achieved significant and important outcomes for third age learning in Hong Kong and even the world.

References

Aberg, P. (2016). Nonformal learning and well-being among older adults: Links between participation in Swedish study circles, feelings of well-being and social aspects of learning. *Educational Gerontology, 42*(6), 411–422.

Boulton-Lewis, G. M., & Buys, L. (2015). Learning choices, older Australians and active ageing. *Educational Gerontology, 41*(11), 757–766.

Brady, E. M., Cardale, A., & Neidy, J. C. (2013). The quest for community in Osher Lifelong Learning Institutes. *Educational Gerontology, 39*(9), 627–639.

Brownie, S. (2014). Older Australian's motivation for university enrollment and their perception of the role of tertiary education in promoting healthy aging: A national cross-sectional study. *Educational Gerontology, 40*(10), 723–736.

Census and Statistics Department. (2017). *Hong Kong population projections 2017–2066*. The Hong Kong Special Administrative Region Government.

Chui, E., & Zhao, X. (2016). Hong Kong. In B. Findsen & M. Formosa (Eds.), *International perspectives on older adult education: Research, policies, practices* (pp. 169–178). Cham: Springer.

Elder Academy. (2018). http://www.elderacademy.org.hk/. The Hong Kong Special Administrative Region Government. Accessed November 12, 2018.

Elderly Commission of Hong Kong. (2001). *Report on healthy ageing*. The Hong Kong Special Administrative Region Government. http://www.elderlycommission.gov.hk. Accessed November 12, 2018.

Findsen, B. (2002). Developing a conceptual framework for understanding older adults and learning. *New Zealand Journal of Adult Learning, 30*(2), 34–52.

Formosa, M. (2014). Four decades of Universities of the Third Age: Past, present, and future. *Ageing & Society, 34*(1), 42–66.

Formosa, M. (2016). Malta. In B. Findsen & M. Formosa (Eds.), *International perspectives on older adult education: Research, policies, practices* (pp. 161–272). Cham, Switzerland: Springer.

Formosa, M. (2019). Active ageing through lifelong learning: The University of the Third Age. In M. Formosa (Ed.), *The University of the Third Age and active ageing: European and Asian-Pacific perspectives* (pp. 3–18). Cham, Switzerland: Springer.

Hong Kong Council of Social Service. (2014). *Lifelong learning at U3A contributes to active aging.* http://www.hkcss.org.hk/cont_detail.asp?type_id=9&content_id=1732. Accessed December 21, 2018.

Hong Kong Council of Social Service. (2016). *U3A envisions greater participation in community affairs.* http://www.hkcss.org.hk/cont_detail.asp?type_id=37&content_id=3219. Accessed December 21, 2018.

Jamieson, A. (2007). Higher education study in later life: What is the point? *Ageing & Society, 27,* 363–384.

Jenkins, A. (2011). Participation in learning and wellbeing among older adults. *International Journal of Lifelong Education, 30*(3), 403–420.

Labour and Welfare Bureau of Hong Kong. (2016). *Response to the Legislative Council on 29 June 2016.* The Hong Kong Special Administrative Region Government. http://www.lwb.gov.hk. Accessed November 12, 2018.

Laslett, P. (1989). *A fresh map of life: The emergence of the Third Age.* London: Weidenfeld and Nicholson.

Lee, P. W., & Chan, K. M. (2002). *Development of lifelong learning among Hong Kong new generative elderly: Abstract of final research report.* Paper presented at the Hong Kong Young Women Christian Association (in Chinese).

Leung, A., Lui, Y. H., & Chi, I. (2006). Later life learning experience among Chinese elderly in Hong Kong. *Gerontology & Geriatrics Education, 26*(2), 1–15.

Lin, Y. Y., & Huang, C. S. (2013). Policies and practices in educational gerontology in Taiwan. *Educational Gerontology, 39*(4), 228–240.

Lun, M. W. A. (2011). Student knowledge and attitudes toward older people and their impact on pursuing aging careers. *Educational Gerontology, 37*(1), 1–11.

Luppi, E. (2009). Education in old age: An exploratory study. *International Journal of Lifelong Education, 28*(2), 241–276.

Manheimer, R. J. (1997). Generations learning together. In K. Brabazon & R. Disch (Eds.), *Intergenerational approaches in aging: Implications for education, policy and practice* (pp. 79–91). New York, US: Haworth Press.

Narushima, M. (2008). More than nickels and dimes: The health benefits of a community-based lifelong learning programme for older adults. *International Journal of Lifelong Education, 27*(6), 673–692.

Newall, N. E. G., & Menec, V. H. (2015). Targeting socially isolated older adults: A process evaluation of the senior centre without walls social and educational program. *Journal of Applied Gerontology, 34*(8), 958–976.

Schuller, T., Preston, J., Hammond, C., Brassett-Grundy, A., & Bynner, J. (2004). *The benefits of learning: The impact of education on health, family life, and social capital.* London and New York: Routledge Falmer.

So, K. M., & Shek, D. T. (2011). Elder lifelong learning, intergenerational solidarity and positive youth development: The case of Hong Kong. *International Journal of Adolescent Medicine and Health, 23*(2), 85–92.

Swindell, R., & Thompson, J. (1995). An international perspective on the University of the Third Age. *Educational Gerontology, 21*(5), 429–447.

Talmage, C. A., Lacher, R. G., Pstross, M., Knopf, R. C., & Burkhart, K. A. (2015). Captivating lifelong learners in the Third Age: Lessons learned from a university-based institute. *Adult Education Quarterly, 65*(3), 232–249.

Tam, M. (2012). Active aging, active learning: Elder learning in Hong Kong. In G. M. Boulton-Lewis & M. Tam (Eds.), *Active aging, active learning: Issues and challenges* (pp. 163–174). Dordrecht: Springer.

Tam, M., Aird, R., Boulton-Lewis, G., & Buys, L. (2016). Ageing and learning as conceptualized by senior adults in two cultures: Hong Kong and Australia. *Current Aging Science, 9*(3), 162–177.

Villar, F., Pinazo, S., Triado, C., Celdran, M., & Sole, C. (2010). Older people's university students in Spain: A comparison of motives and benefits between two models. *Ageing & Society, 30*(8), 1357–1372.

Withnall, A. (2010). *Improving learning in later life*. London, United Kingdom: Routledge.

Zhang, W. Y., & Ha, S. (2001). *An investigation into the learning attitude, motivation and preferences of the older adults in Hong Kong*. Paper presented at the 15th Annual Conference of the Asian Association of Open Universities, India Gandhi National Open University, New Delhi, India.

Maureen Tam is Professor at the Education University of Hong Kong (EdUHK), Co-Director of the Centre for Lifelong Learning Research and Development, as well as Head of the EdUHK Elder Academy. Previously, she was Dean of the Community College and Further Education, and Director of the Teaching and Learning Centre of Lingnan University responsible for teaching and learning development, quality assurance, assessment of university effectiveness and student experiences. Her research interests are wide-ranging and cover all aspects of older adult education, lifelong learning, professional and vocational education, quality assurance, outcomes-based education, and teaching, learning and assessment in higher education.

Chapter 15
The University of the Third Age in Lebanon: Challenges, Opportunities and Prospects

Maya Abi Chahine and Abla Mehio Sibai

Introduction

Located in the Middle East, Lebanon has a rich history dating back more than 6000 years. Its population is close to 4.5 million and features diverse cultures and religions that include 18 different sects. Unlike some of its neighbouring countries, Lebanon has only recently discovered oil and gas and is classified as a middle-income country. Driven by economic development and advancement in the healthcare system and modernity, the past four decades have witnessed increasing trends in population ageing in Lebanon (United Nations, 2017). While the crude birth rate decreased from 28.8 in 1980 to 15 in 2015, the crude death rate decreased from 7.2 to 4.6 per thousand in the same period (ibid.). During the same period, the fertility rate dropped from 3.75 in 1980 to 1.72 in 2015, and life expectancy at birth increased from 68.4 to 79.8 (ibid.).

The above population trends have led to the rectangularisation of the population pyramid and an increase in the size of the older population. Indeed, adults aged 65 and over constitute 8.2% of the population and are projected to reach 10.20% in 2025 and 23% in 2050—the highest percentage in the Arab world (United Nations, 2017). The population of those 80 and above is projected to quadruple, from 0.9 to 4.3% over the 2025–2050 period. Lebanon is also experiencing a feminisation of the older population with a male/female ratio of 0.86 for those aged 85 years and above. An additional factor playing a role in the 'greying' of the population is two-way migration. Emigration among the youth is caused by economic recession, conflict and turmoil, which push young adults to seek better opportunities outside the country, while reverse migration witnesses an increasing number of older adults returning to

M. A. Chahine (✉) · A. M. Sibai
American University of Beirut, Beirut, Lebanon
e-mail: ma271@aub.edu.lb

A. M. Sibai
e-mail: am00@aub.edu.lb

© Springer Nature Switzerland AG 2019
M. Formosa (ed.), *The University of the Third Age and Active Ageing*, International Perspectives on Aging 23, https://doi.org/10.1007/978-3-030-21515-6_15

Lebanon after retirement (Sibai, Rizk, & Kronfol, 2014). In the dawning of the new millennium, some 20% of Lebanon's population has migrated due to political unrest (World Bank, 2008). Leading the Arab world in terms of population ageing, Lebanon has also led the region when it comes to Universities of the Third Age (U3A) with the first, and only, Lebanese U3A being established in 2010.

This chapter presents the authors' experience regarding the U3A in Lebanon and presents our findings in six sections. The first highlights the genesis of the U3A movement in Lebanon, highlighting the U3A's key concepts and features. The second examines the community and presents the profile of the people who benefit most from it. The third section provides details on the curriculum, teaching and learning modalities. The contributions of the programme to active ageing in Lebanon as well as its impact on the well-being of seniors are the subject of Section four and the challenges faced are discussed in Section five. Finally, this chapter ends with a vision for the future and the potential of the U3A and older adults in general, and in Lebanon, in particular.

Genesis

The genesis and development of the U3A movement in Lebanon can only be understood within the context of the country's literacy levels and the social and economic profile of its older population. Lebanon has played a major role in the Arab world in terms of education. Since the mid-nineteenth century, Lebanon's schools and universities have represented a beacon of learning. Lebanon has the lowest level of illiteracy in the region (28.2%) compared to, for example, Oman which stood at 72.3% (Sibai & Kronfol, 2011). However, literacy levels are not consistent across age brackets. For instance, among those aged 80 and above, only 36.1% have attended school; there are also gender-related discrepancies, with older women being twice as illiterate as older men (45.8% vs. 24.8%, respectively) (ibid.). It is nevertheless positive to note that as schooling continues to improve, this gender gap is expected to close. Moreover, owing to a poor pension system, around 26% of men continue working past retirement age (64 years) of which 11% are over 80 years old; older women's participation in the labour market reaches only 2.8%, with many performing critical unpaid services mainly in agriculture or as caregivers (ibid.). While divorce is very rare among older generations, widowhood is more common among women than among men (50.3 and 13.9%, respectively) due to a greater tendency for older men to remarry following divorce or widowhood (ibid.). This leads to 17.4% of women living alone compared to 6.2% of men, a ratio similar to Western countries and the highest in the Arab world (Tohme, Yount, Yassine, Shideed, & Sibai, 2011). However, while 'older persons living alone defy the customary Arab arrangement of intergenerational co-residence and support' (Kronfol, Rizk, & Sibai, 2016: 209), studies examining the drivers behind living arrangements in Lebanon are lacking.

Demographic transition is occurring much faster in Lebanon than it did in the Western European and Northern American countries, with Lebanese society remaining utterly unprepared to face the consequences of population ageing. Although Lebanon's older cohorts are living healthier and experiencing longer lives than previous generations, with many older persons remaining relatively healthy and independent far into their later years, the opportunities provided to them in terms of social engagement and intellectual stimulation remain limited. It was in direct response to such a state of affairs that, in 2008, two professors of Public Health at the American University of Beirut (AUB), Cynthia Myntti and Abla Mehio Sibai, launched a feasibility study to explore the potential for a U3A in Lebanon.[1] AUB was a natural home to the first and only U3A in the region given the leading role in education that it has played over its 150-year life, since none of the Arab states have lifelong learning policies and only five have lifelong learning programmes. These programmes target mainly professional development issues, while other programmes targeting older adults are concerned mainly with literacy. Indeed, to the authors' knowledge Lebanon remains the only country in the Arab world to host a U3A, namely the University for Seniors (UfS) at AUB (see https://www.aub.edu.lb/rep/cec/uni_seniors/Pages/main.aspx). Similar to other U3As in the world, three key catalysts drove the emergence for such a programme: the worldwide increase in the older adult population, the feminisation of later life and a general improvement in the health status and educational level of older persons (Formosa, 2002, 2005, 2012a). The fact that the Madrid International Plan of Action on Ageing (MIPAA) has long stressed the importance of education 'for an active and fulfilling life' (United Nations, 2002: 26) was central to the interest and determination of AUB in hosting the U3A.

The creation of AUB's UfS followed a bottom-up approach after an extensive two-year feasibility study. The study consisted of focus group discussions with seniors from different backgrounds, a survey of older AUB alumni, a questionnaire sent to well-established university-based centres in the USA and Canada, and visits to third age learning programmes in Harvard, the University of Minnesota and New York University. This participatory approach was in contrast with U3As following the French model, such as the U3A in Malta, which was developed in a top-down approach by the university academicians without consulting the older persons themselves (Formosa, 2012a, 2014, 2016). This approach allowed older learners at AUB to be able to input into the preferred title of the programme, the topics, length of terms, number of sessions per week and their length, timing and membership fees among other logistic matters. The feasibility study also demonstrated that a U3A that stresses on community building rather than skill development for career enhancement would be more enthusiastically received in the country. The UfS was born in the Spring of 2010 and was housed in AUB's Continuing Education Center. The programme's vision is to create a new, positive, approach to ageing in Beirut, Lebanon and the Middle East, so that older adults are given an opportunity to remain intellectually and socially engaged, learn new things and be active contributors to their communities. The UfS's mission is to provide older adults with educational and

---

[1] The American University of Beirut is a private university, ranked second in the Arab region as per QS University Rankings, Arab Region (QS University Rankings, n.d.).

cultural opportunities in a sociable environment. The programme offers a variety of activities: study groups, lectures, cultural travel programmes and intergenerational activities with current AUB students (American University of Beirut, n.d.). The UfS is open to women and men over the age of 50. This age was opted for, rather than 60 or 65 and above, as is the case in many other U3As around the world, because the idea of learning in later life was completely alien to Lebanese society in 2010, especially among the oldest old. In addition, enrolling adults at 50 would promote an early interest in the programme and increased enrolment as potential participants grow older.

Traditionally, two models have shaped U3As around the world. Whilst the British U3A model 'emphasizes informal, autonomous self-help groups…teach their peers', the French model is much more academic and top-down (Villar & Celdrán, 2012: 667). Formosa (2014) identified four other models—the 'culturally hybrid', 'French-speaking North American', 'South American' and 'Chinese' U3A models—in addition to U3As which operate through the online medium. Lebanon's U3A draws on the experience of 'institutes for learning in retirement' in North America and the Universities of the Third Age from both the French and the British models, creating a unique hybrid that is adapted to both context and culture. The programme combines elements from both models: it is based in a university but operates on a participatory approach whereby members can influence the programme's strategy by contributing to its by-laws and policies, shaping the curriculum, proposing lecturers for recruitment, contributing towards intergenerational solidarity connections and strengthening the community spirit. There are no paid teachers, and the academic body is drawn from university professors, as well as the members themselves or individuals from the community, who are ready to share their expertise and passion on a voluntary basis. UfS is run by a Programme Manager and a Programme Assistant, who constitute the only members of staff who receive an official remuneration. The programme is for and by the seniors as UfS members take an active part in its governance through membership in three committees. These committees (curriculum, social and institutional) play a vital role in running and developing the programme. Members also volunteer to coordinate trips, social media posts and help in classes and in the office. It is estimated that volunteering saves UfS close to $300,000 per year.

In retrospect, the UfS programme draws its uniqueness from three principles: first, peer learning, as participants are expected to share knowledge and learn from each other, and there are no paid teachers. Seniors with a passion and expertise in a certain subject volunteer to facilitate a study group or give a lecture; second, community building, as one joins a term rather than enrolling for a specific course, with social events being organised throughout the year; and finally, intergenerational solidarity connections, as seniors are connected to the AUB student body through intergenerational academic and extra-curricular activities, both on and off campus.

Membership

The UfS Membership has consistently increased since inception: from 57 members in Spring 2010 to 286 in Spring 2017 and from 315 subscribers to the emailing list to 1688 (Fig. 15.1). The programme, given its resources and facilities, cannot accommodate more than 300 members per term. This helps preserve its personalised, community spirit.

Over the past 7 years, the UfS has attracted 816 unique members with 2422 enrolments over 14 terms. Details on our members are presented below with an attempt to draw similarities with U3A programmes in other countries (see Formosa, 2007, 2010, 2012b, 2014) (Fig. 15.2).

Age. While the UfS programme is open to anyone aged 50 and above, 69% of members are seniors aged over 60, with sexagenarians making up the predominant age group since 2010 (40%). Around 4% are octogenarians.

Gender. Women outnumber men across all groups (between 65 and 84%). Although male membership is increasing in number, it is not increasing by overall percentage. This is due to the fact that men continue working long after retirement age in Lebanon. While they could feel anxious by joining a programme dominated by women, one cannot ignore the psychological factor in Arab culture whereby men are pressured to show that they 'know it all'.

Moreover, the subjects offered tend to be more appealing to women than men, although in recent years we have included more subjects concerning politics and economics, which are often requested by men. At the same time, Lebanon is no exception to the phenomenon of feminisation of ageing. The high number of women living alone, in addition to higher percentages of women who have never attended college, makes UfS a very attractive proposition to older women. Women are often

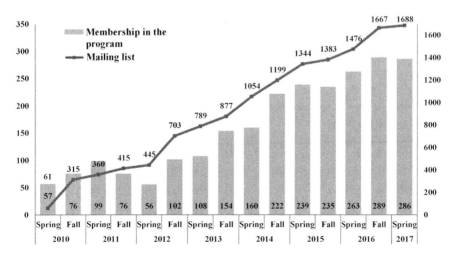

Fig. 15.1 Membership in the programme

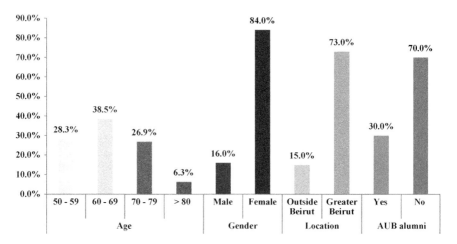

Fig. 15.2 Characteristics of the UfS members in Spring 2017

heard emphasising how UfS has offered them a safe environment where, for the first time in their lives, they feel that they are doing something for 'themselves and not for their spouses or their children'.

Social class. As with most U3As around the world, the UfS predominantly attracts middle-class seniors. However, the nominal membership fee of $200 is relatively low by Lebanese standards and, thus, allows for more diversity than is normally seen in a segregated society like Lebanon (the minimum wage in Lebanon is $450 per month). Organising classes at various times of the day (between 10.00 and 19.00) and offering a third of the curriculum in Arabic have improved the membership diversity. Although it appears that more older persons from the low-middle class are joining UfS, there are no available data confirming this possible trend, because for equity purposes socio-economic information is not requested in the registration form. One proxy indicator regarding the social diversity of membership is the proportion of our members who are AUB alumni. This proportion dropped from almost 60% in Spring 2010 to 30% in Spring 2017. Although UfS started as a programme that attracted mainly AUB alumni and seniors from more wealthy neighbourhoods, more members are hailing from the Greater Beirut area, with almost 20% coming from distant regions. However, the pervasive traffic problems in Lebanon, a lack of adequate public transport and accessible parking render attempts to reaching older persons residing in distant communities very difficult.

Third agers. Seniors enrolling in the programme are mostly physically fit and independent. Yet the programme does not shy away from taking responsibility for the physically challenged and others with mild cognitive impairment, albeit only a very small number (17 cases noted in the Spring 2017 term). A shuttle service is provided to those who cannot walk to class and all classes are held in venues that are easily accessible to seniors.

Curriculum, Teaching and Learning

As a hybrid programme combining elements from both the French and British models, Lebanon's U3A mostly falls in the humanist camp—as do the offerings of most U3As worldwide—although as discussed in the previous chapter, it also contains elements from critical education gerontology. The number of offerings at the UfS has been increasing exponentially. In Spring 2010, the UfS offered 26 contact hours, increasing gradually to 80 in Spring 2017. UfS's curriculum draws mostly on the liberal arts. In Spring 2017, 32 liberal arts sessions were offered, twice as many as other topics, grouped under health, technology and special interest groups. The programme changes every term and the topics are always varied, covering a wide array of information and skills ranging from philosophy, history, politics, arts, health, literature, technology, aromatherapy, karate, bridge, yoga, folk dancing, educational trips, etc. Aware of the need to diversify U3As' programmes by offering non-liberal and non-heath related topics (Formosa, 2014), UfS has offered a financial literacy series for the past two terms and intends to make this a regular offering (Fig. 15.3).

Although some U3As have led their members to associate themselves more with the middle age cohorts rather than perceiving themselves as third agers (Formosa, 2012c), the UfS' vision is to reflect a positive image of ageing and utilise its curriculum to anchor its members in a third age reality. Indeed, the curriculum focuses on age-related topics such as the ageing of the face, palliative care, end-of-life decisions and mental health in old age. The programme constantly highlights the members' value and rich experience by offering memoir-writing and oral history classes. A recent initiative, *Seniors to Seniors*, wherein UfS members conducted mock job interviews with AUB senior year students offered the latter the opportunity to be mentored by select, experienced seniors. Special focus is also given to classes on

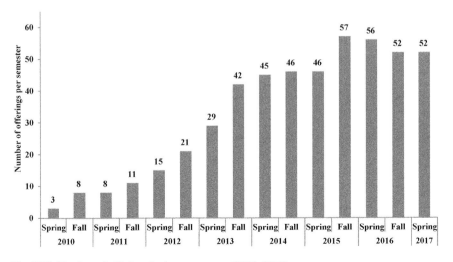

Fig. 15.3 Number of offerings in the programme (2010–2017)

information and communication technology, which are offered by the regular AUB student body. This forges intergenerational connections and is in line with the *Madrid International Action Plan on Ageing*'s (United Nations, 2002: 27) proposal to 'provide opportunities within educational programmes for the exchange of knowledge and experience between generations, including the use of new technologies'. These classes have attracted close to 750 registrations to date.

Teaching methodologies are diverse, and each term the programme offers a combination of lectures, study groups (hands-on classes), special interest groups, book clubs, cinema clubs and educational trips. These are offered by a body of volunteer 'teachers' comprising 270 volunteers to date. The teaching body is very diverse: 65 professors, 25 UfS members, 95 seniors from the community and 85 others who are either AUB students or non-seniors from the community sharing their expertise and passion. As close to 50% of our teachers are females, this gender balance is an achievement considering that U3As generally have a predominantly male teaching body (Formosa, 2014). While members participate by suggesting topics and lecturers, or by delivering classes themselves, they remain recipients of the material delivered by the lecturers. Members gain knowledge and skills and enhance their health and financial literacy. This is contributing to their well-being and to changing behaviour on the personal level (detailed in the impact section). In certain instances, UfS classes have driven members to act and make a change in their community. Following a lecture about the 2016 municipal elections, some members worked with a new movement lobbying for change in the country. Other members became involved in a road safety awareness project after attending a talk on the issue. This interest in playing a greater role as agents of change is growing. In a small UfS town hall meeting in January 2017, members asked UfS to facilitate their involvement in activities for the advancement of seniors' rights and other issues that affect their communities.

Impact of the Programme

The contributions and impact of U3As have been studied and reported, notably by Villar and Celdrán (2012) and Formosa (2010, 2014). The studies cite, among other benefits, an enhancement in the quality of life, improved health, psychology and social life, democratising lifelong learning, including later life learning in policies and more. Seven years after its establishment, UfS is a key contributor to enabling older persons in Lebanon to practice active ageing, by allowing its members to have greater 'participation in social, economic, cultural, spiritual and civic affairs' (World Health Organization, 2002: 12). Its impact on the quality of life and well-being is also visible through abundant anecdotal information and studies. It is a feather in its cap that the UfS was recently selected by students from Oulu University in Finland and from Aarhus University in Denmark for their Masters' thesis in evaluating the impact of U3As on the lives of its members. One paper has already been published (Hachem & Vuopala, 2016). Moreover, the UfS was also recently selected by the World Health Organisation and the RAND Institute as one of the 10 most innovative community-based initiatives in middle-income countries that seek to empower older people and promote social cohesion and inclusiveness, and research is underway to examine the impact of the UfS on the quality of life of its members.

On possible impact that strikes the authors is the change in 'ageist' attitudes in the participant's lives. Older adults are highly respected in Lebanese culture, but they are also often described as frail and dependent individuals. Some of the 'ageist' terms used are words like 'ajouz' coming from 'ajaz' which means 'main disability', 'khitiar' which means physically and mentally impaired and 'sin el ya's' which means the 'age of despair'. The image portrayed by the UfS on campus, and in the country, through media coverage in local and regional outlets, is positive, dynamic, engaged and empowered. The programme is enhancing a sense of empowerment and engagement, pushing seniors to make a change within, and sometimes around, themselves. Various facets of the potential impact of the UfS are as follows:

- Due to the civil war (1975–1990) and the repercussions of sociopolitical changes and conflicts in neighbouring countries, Lebanon is starkly divided along religious and class lines with very few public spaces where people from different backgrounds can meet on an equal footing. As the programme projects an inclusive character, UfS is becoming one such possible public space.
- Public health practitioners classify UfS as a primary health prevention, which promotes physical, emotional and social health. Geriatricians and psychologists are both encouraging their clients to become members, as they observe a vibrant community spirit among members. The UfS thus functions to improve the members' levels of social capital.
- Children of members report the positive change in their parents' lives and the family as a whole, with one claiming to the authors: 'my mother has been ill for the past two years. UfS…is giving her reason to carry on'.
- The impact on women is especially significant as they are offered the opportunity to break free from their traditional social roles. One female member disclosed to the authors: 'I no longer feel neglected over family dinners as I have many topics to share and discuss now'.
- The impact on physical capital is also reported in the term evaluations that members voluntarily complete, as one member wrote: 'I like to walk, so I had the chance to practice this sport in a beautiful place'.

Hachem and Vuopala (2016) studied 461 UfS members who registered in the Spring and Fall of 2016, with as much as 143 members completed an end-of-term evaluation. The benefits reported by the members were grouped into cognitive, social and psychological responses. Cognitive benefits were the most reported (62%), with members stressing how enriching the new information and acquired skills are: 'it is enriching to me that I learned new skills, like how to use the iPad, in addition to new medical information'. Social benefits were reported in 58% of the responses as members emphasised the community spirit, making new relationships and social learning: 'the UfS boosted my morale, spirit and sense of belonging so much that I feel proud and confident being part of it (UfS) and…gained a large, supportive family'. Finally, psychological benefits appeared in 48% of the responses, highlighting feelings of happiness, positivity and self-confidence: 'I am more relaxed, and the program keeps me excited throughout the week…it gave me more self-confidence and satisfaction'. Members also stated that the programme helped them cope with difficult changes in their lives, such as losing family members or retiring: 'when I quit my job, my life was empty and meaningless, but with [the programme] my life changed and is more filled with joy, education and interaction'. Hachem and Vuopala (2016) concluded that these findings are in line with international studies, and hence, it results that the benefits of U3A are universal despite the unique nature of each programme.

Challenges Faced by the U3A

Formosa (2014) depicted a number of challenges faced by U3As citing, among others, the lack of suitable venues and sufficient volunteers in the British model, and that U3As following the French model are mainly located in cities, are characterised by a membership body that is far from inclusive (gender, class, ethnicity and age) and lacking intergenerational connections. The UfS is also categorised by such lacunae, for example by attracting less older men and older persons from remote areas and being relatively sparse of intergenerational activities. While the programme is trying to accommodate moderately physically and cognitively impaired older persons, such attention and inclusive agenda become impossible to maintain when their health deteriorates past a certain critical stage. While it is noteworthy that even European policies overlook the learning needs of frail and vulnerable older persons (Formosa, 2012b), it is hoped that Lebanon will be able to include both third and fourth agers when drafting late-life learning policies.

The UfS always struggled with finance and human resources, and the programme relies on membership revenues and sporadic donations which do not guarantee financial sustainability, especially in a country that suffers from political unrest. Indeed, Hachem and Vuopala (2016) located three categories of challenges during their study of older adult learning in Lebanon: institutional, dispositional and situational. Institutional challenges were related to the curriculum or physical infrastructure, as some members requested more sessions on specific topics in order to develop skills or requested that the capacity of study groups be increased, and others complained about 'repetitive, irrelevant and unnecessary questions'. It was noted that newcomers always need a term or two to fully grasp the rational and ethos of the programme, while long-standing members complained about breaches of class protocol, especially with the rising number of newcomers every term. While one potential solution is holding an orientation session and assigning mentors to help new members, it is not clear as how problems related to inadequate infrastructure—as classrooms and venues often have poor audio-visual equipment and uncomfortable chairs—may be resolved without an increase in financial resources. While dispositional challenges were largely related to the level of difficulty and language barriers (some requesting more lectures in Arabic, while others preferred lectures conducted through the

English medium), situational challenges revolved mostly around class scheduling as members complained that classes were held either too early in the morning or too late in the evening.

Conclusion

Lebanon's U3A has always been a staunch supporter of humanist philosophy, and both the academic team and member committees are now exploring ways how to turn it into a flagship for change as inspired by Inglis' (1997) ideas of empowerment and emancipation. In a country that lacks age-related policies, where older adults are living longer and healthier, are better educated, aspire to remain active as they age and on the basis of the UfS's long-term successes, discussions are presently underway with the AUB administration to create the first *Programme of Excellence on Ageing* in Lebanon and the Arab world. The programme, in which UfS would be embedded, will work on advancing research, education and advocacy to place older adult issues on the national agenda. Creating an activist group from UfS members (and others) is another strategic point under consideration. At the same time, members are considering outreach activities where they would offer technical support to local businesses and non-governmental organisations which constitutes one of the pillars of critical educational gerontology pillars whereby 'both teachers and learners need to extend their work outside the educational setting, and connect with what is going on in the public sphere' (Formosa, 2011: 327). However, this platform for ageing issues cannot be isolated from regional and international efforts in advocacy. To this effect, the UfS has been working with various United Nations agencies—International Institute on Ageing (United Nations—Malta), World Health Organization, United Nations Economic and Social Commission for Western Asia, United Nations Populations Fund—and research institutes such as RAND Europe. Collaboration is also underway with other non-governmental organisations such as HelpAge International with whom the UfS is collaborating towards the organisation of a regional conference on ageing to highlight regional issues and create a plan of action. Given UfS's uniqueness, potential and impact, requests to replicate the programme have been received from Lebanon and abroad (namely Palestine and Bahrain), and feasibility studies are currently underway. This would allow the UfS to have a larger and more inclusive membership, especially in remote areas. Such a goal could also be achieved through online learning (which would allow the provision of such a service to more frail and vulnerable older persons, the fourth agers, who are generally unable to leave their residence). As a first step, live Skype conferencing was successfully launched in Spring 2017. Lebanon's late entry into the U3A movement offers a unique opportunity to harness and build on four decades of international experience to meaningfully face the region's daunting ageing challenges.

References

Formosa, M. (2002). Critical gerogogy: Developing practical possibilities for critical educational gerontology. *Education and Ageing, 17*(1), 73–86.

Formosa, M. (2005). Feminism and critical educational gerontology: An agenda for good practice. *Ageing International, 30*(4), 396–411.

Formosa, M. (2007). A Bourdieusian interpretation of the University of the Third Age in Malta. *Journal of Maltese Education Research, 4*(2), 1–16.

Formosa, M. (2010). Universities of the Third Age: A rationale for transformative education in later life. *Journal of Transformative Education, 8*(3), 197–219.

Formosa, M. (2011). Critical educational gerontology: A third statement of first principles. *International Journal of Education and Ageing, 2*(1), 317–332.

Formosa, M. (2012a). Education for older adults in Malta: Current trends and future visions. *International Review of Education, 58*(2), 271–292.

Formosa, M. (2012b). Education and older adults at the University of the Third Age. *Educational Gerontology, 38*(2), 114–126.

Formosa, M. (2012c). European Union policy on older adult learning: A critical commentary. *Journal of Aging and Social Policy, 24*(4), 384–399.

Formosa, M. (2014). Four decades of Universities of the Third Age: Past, present, and future. *Ageing & Society, 34*(1), 42–66.

Formosa, M. (2016). Malta. In B. Findsen & M. Formosa (Eds.), *International perspectives on older adult education: Research, policies, practices* (pp. 161–272). Cham, Switzerland: Springer.

Hachem, H., & Vuopala, E. (2016). Older adults, in Lebanon, committed to learning: Contextualizing the challenges and the benefits of their learning experience. *Educational Gerontology, 42*(10), 686–697.

Inglis, T. (1997). Empowerment and emancipation. *Adult Education Quarterly, 48*(1), 3–17.

Kronfol, N., Rizk, A., & Sibai, A. M. (2016). Ageing and intergenerational family ties in the Arab region. *Eastern Mediterranean Health Journal, 21*(11), 835–843.

QS University Rankings. (n.d.). *QS University rankings Arab region 2016*. https://www.topuniversities.com/university-rankings/arab-region-university-rankings/2016. Accessed March 15, 2018.

Sibai, A., & Kronfol, N. (2011). *Older population in Lebanon: Facts and prospects*. Country Profile, Center for Studies on Aging and United Nations Population Fund. http://www.csa.org.lb/cms/assets/archives/2011/13%20country%20profile%20elderly%20population%20in%20lebanon%20facts%20and%20prospects%20english.pdf. Accessed August 10, 2017.

Sibai, A. M., Rizk, A., & Kronfol, N. M. (2014). *Ageing in the Arab region: Trends, implications and policy options*. Center for Studies on Aging, United Nations Social and Economic Commission for Europe, and United Nations Population Fund. http://csa.org.lb/cms/assets/csa%20publications/unfpa%20. Accessed August 10, 2017.

Tohme, R. A., Yount, K. M., Yassine, S., Shideed, O., & Sibai, A. M. (2011). Socioeconomic resources and living arrangements of older adults in Lebanon: Who chooses to live alone? *Ageing & Society, 31*(1), 1–17.

United Nations. (2002). *Madrid international plan of action on ageing*. New York: United Nations.

United Nations. (2017). *World population prospects: The 2017 revision, key findings and advance tables*. Working Paper No. ESA/P/WP/248. https://esa.un.org/unpd/wpp/Publications/Files/WPP2017_KeyFindings.pdf. Accessed August 11, 2017.

Villar, F., & Celdrán, M. (2012). Generativity in older age: A challenge for universities of the third age (U3A). *Educational Gerontology, 38*(10), 666–677.

World Bank. (2008). *The migration and remittances fact book 2008*. https://openknowledge.worldbank.org/bitstream/handle/10986/6383/429130PUB0Migr101OFFICIAL0USE0ONLY1.pdf?sequence=1&isAllowed=y. Accessed August 20, 2017.

World Health Organization. (2002). *Active ageing: A policy framework*. http://apps.who.int/iris/bitstream/10665/67215/1/WHO_NMH_NPH_02.8.pdf. Accessed March 15, 2018.

Maya Abi Chahine is Programme Manager of the University for Seniors at the American University of Beirut, the first lifelong learning initiative in the Middle East addressed exclusively to older adults. She has 20 years of experience in education management, community development and outreach, and a solid knowledge and experience in Ageing issues and Third Age Learning. She has worked on several studies about the impact of lifelong learning on seniors' health and well-being in Lebanon. She holds an MA in public policy and ageing from King's College London (2018) and an MA in political science and public Administration (1998) from Saint Joseph University, Lebanon.

Abla Mehio Sibai is Professor of Epidemiology and Population Health at the Faculty of Health Sciences, American University of Beirut and also the co-founder of the AUB University for Seniors program and the Center for Studies on Aging in Lebanon. She has led several studies on ageing in Lebanon and the Arab region and is frequently consulted by Ministries in Lebanon, WHO, UNFPA and ESCWA on older adult issues. Abla is the author of over 200 articles in prestigious journals, book chapters and scientific reports. She holds a Ph.D. from the London School of Hygiene of Tropical Medicine.

Chapter 16
Moving the Needle on the University of Third Age in Malaysia: Recent Developments and Prospects

Tengku Aizan Hamid, Noor Syamilah Zakaria, Nur Aira Abd Rahim, Sen Tyng Chai and Siti Aisyah Nor Akahbar

Introduction

Malaysia is an upper-middle income country with a gross domestic product (GDP) of United States Dollars (USD) 296.4 billion in 2015. Its gross national income (GNI) per capita ranks 3rd in Southeast Asia after Brunei and Singapore and 44th in the world (World Bank, 2017). In the population projections released by the Department of Statistics Malaysia, 2.9 million or 9.2% of the total 31.4 million population was aged 60 years or over in 2015—the official definition of an older person in the country (Ministry of Women, Family & Community Development, 2011; Department of Statistics Malaysia, 2016). Between 2020 and 2039, the proportion of older Malaysians aged 65 years or over is expected to double from 7.2 to 14.1%, a growth

T. A. Hamid
Malaysian Research Institute on Ageing, Universiti Putra Malaysia, Serdang, Malaysia
e-mail: aizan@upm.edu.my

N. S. Zakaria
Department of Counsellor Education and Counselling Psychology, Faculty of Educational Studies, Universiti Putra Malaysia, Serdang, Malaysia
e-mail: syamilah@upm.edu.my

N. A. A. Rahim
Department of Professional Development and Continuing Education, Faculty of Educational Studies, Universiti Putra Malaysia, Serdang, Malaysia
e-mail: nuraira@upm.edu.my

S. T. Chai (✉)
Laboratory of Social Gerontology, Malaysian Research Institute on Ageing, Universiti Putra Malaysia, Serdang, Malaysia
e-mail: chez1978@gmail.com; chai@upm.edu.my

S. A. N. Akahbar
Laboratory of Medical Gerontology, Malaysian Research Institute on Ageing, Universiti Putra Malaysia, Serdang, Malaysia
e-mail: boocomei@yahoo.com

© Springer Nature Switzerland AG 2019
M. Formosa (ed.), *The University of the Third Age and Active Ageing*, International Perspectives on Aging 23, https://doi.org/10.1007/978-3-030-21515-6_16

of 242% in absolute numbers, from 2.4 million to 5.8 million persons. Median age of the population has increased from 17 years in 1970 to 30 years in 2020. By any measure, Malaysia is a rapidly ageing society, and this poses significant opportunities and challenges for the multi-ethnic democracy in an open, modern economy. Malaysia's swift demographic transition and economic transformation have gone hand in hand in recent decades. Since gaining independence in 1957 and after the formation of Malaysia in 1963, the country underwent rapid industrialisation and maintained a high, export-oriented growth where the GDP grew by an average of 6.3% per annum between 1970 and 2015. GNI per capita (Atlas method, current USD) grew exponentially from USD370 to USD10,450 in the same period. Population growth, however, slowed considerably as the total fertility rate (TFR), or the average number of children born per woman between 15 and 49 years old, dropped from 4.9 in 1970 to reach below replacement levels in 2015 (Department of Statistics Malaysia, 2017a). Average life expectancy at birth has grown from 61.6 years for males and 65.6 years for females in 1970 to 72.5 and 77.1 years, respectively, in 2015 (ibid., 2015, 2017b). With the added years to life, successive generations of older Malaysians are living longer in retirement, although the sociocultural landscape has changed dramatically as the family system and its values are evolving in tandem with modernisation and urbanisation trends. This chapter is organised into three sections, first providing an overview of the country situation on active ageing and lifelong learning, followed by a description of the University of the Third Age (U3A) programmes in Malaysia, and finally, a discussion on the future of U3A and other similar movements nationally.

Active Ageing and Lifelong Learning in Malaysia

The development of national policies for older persons in Malaysia has been extensively covered in past literature (Hamid & Chai, 2013; Hamid & Yahaya, 2008; Ong, Phillips & Hamid, 2009; Tey et al., 2016). It is important to note that the United Nations Principles for Older Persons adopted in 1991 during the United Nations General Assembly (Resolution 46/91) laid the foundation for the first National Policy for the Elderly (NPE) (Ministry of National Unity and Social Development, 1999). Rooted in welfarism and provisions for protection and care, the NPE first underscored the dignity, potential and well-being of the older population. Although the second objective of the NPE was to "improve the potential of the elderly [sic] so that they continue to be active and productive in national development…" (NPE, 1995), the government's emphasis throughout the 1990s was mostly on the family. Themes for the National Day of Older Persons (NDOP), first celebrated in 1992, would remain a variant of caring or loving families until the turn of the new millennium. It was only in 2000 when the NDOP theme was changed to "Active Elderly, Productive Lives".

In the 9th Malaysia Plan (2006–2010) (Economic Planning Unit, 2005: 311), the section on "Older Persons" noted the progress achieved, highlighting a shift of programmes "from a welfare approach to a development approach to ensure active and productive ageing", where phrases such as lifelong learning and ICT were first associated with older persons in the five-year national development plans. The government promised to undertake measures "to provide for an environment for older persons to remain healthy, active and secure while being able to age with dignity and respect as well as leading independent and fulfilling lives as integral members of their families, communities and country" (ibid., 316–317). This progressive focus persisted in the 10th Malaysia Plan (2011–2015) (ibid., 2010) where the government promoted greater inclusivity by dedicating a section on "Supporting older persons to lead productive and fulfilling societal roles". The Plan promised "to provide a conducive environment for older persons to remain healthy, active and secure" (ibid., 185). In the latest 11th Malaysia Plan (2016–2020) (EPU, 2015), the strategy to enhance the living environment for older persons specifically highlighted the need for active ageing, and it noted that

> ...the University of the Third Age (U3A) programme will be expanded to provide opportunities for the elderly to continuously learn and acquire knowledge and skills in collaboration with universities, community colleges and Older Person Activity Centres (PAWE). In addition, NGOs will also be encouraged to implement self-help based learning programmes.
>
> Economic Planning Unit (2015: 3–27)

As how these promises translate into concrete actions on the ground will now be discussed at length in this section. As noted by Ibrahim, Zakaria, Hamid, Chai, (2017), lifelong learning in Malaysia is still focused primarily on skills retraining and vocational education. As per the *Blueprint on Enculturation of Lifelong Learning for Malaysia* (2011–2020) (Ministry of Higher Education, 2011: 60), community colleges are the designated entity to offer programmes relevant to the needs of the local population, assisting the poorest segments of society such as the underprivileged, the disabled and senior citizens "to enhance their communication and computer skills". However, this was not a specific suggestion made in isolation. As early as 2003, the government, through the Demonstrator Application Grant Scheme (DAGS), provided financing for computer literacy programmes to bridge the digital divide among older persons. It was widely recognised that older Malaysians could be further marginalised if they do not possess the necessary information technology literacy and skills to survive in the Digital Age, but those were the days when smartphones are still a novelty and most people accessed the World Wide Web through computers.

It is noteworthy that older Malaysians had their own cultural groups and social circles, long prior to any government policies and programmes on ageing. Merriam and Mohamad (2000) found that older adult learning in Malaysia, in particularly among the Malay older persons, is non-formal and experiential, is communal and much of it is religious or spiritual in orientation. On the other hand, the Senior Citizens Club of Perak just celebrated its 40th anniversary in 2017 with a dinner and dance at a hotel. Throughout the country, more common in towns and cities with a sizable Chinese population, there are numerous senior citizens' clubs that serve as a gathering spot for older persons to socialise and enjoy themselves—dancing, singing and travelling. These social groups reflect the leisurely pursuits of an ageing middle-class population, although the establishment of a National Council of Senior

Citizens Organisations Malaysia in 1990 quickly cemented this federation of over 40 affiliates as a representative voice for older persons. When the government began to promote active and productive ageing lifestyles, lifelong learning activities for older persons became increasingly popular. For example, the Senior Citizens' Club at YMCA Kuala Lumpur started a Young-at-Heart programme in 2006 where older persons could participate in learning programmes on human development and ageing, health and personal care, as well as other craft or hobbyist electives. While lifelong learning for older persons might conjure up images of seniors in computer classes, many retirees have always participated in activities that cultivate their abilities to paint, gardening or playing musical instruments.

A quick search at the Registrar of Societies' online database yielded nearly 400 senior citizen clubs/health clubs, retiree associations and pensioner groups, even though many senior citizen or retiree groups do not register themselves. As part of its strategy under the *Action Plan for the National Policy for the Elderly* (1997), the Ministry of Health encouraged the set-up of Senior Citizen Health Clubs (*Kelab Kesihatan Warga Emas*) at each and every one of the government's primary health clinics. In its 2009 Annual Report, the Ministry of Health (2009) reported 240 such clubs being established nationwide. In 2012, the Department of Social Welfare began setting up Older Person Activity Centres (*Pusat Aktiviti Warga Emas*) by converting existing Elderly Day Care Centres (*Pusat Jagaan Harian Warga Tua*) or establishing new ones. In the same year, a major recreational centre for older persons in Ampang built by the Aged United to Organise Rest and Recreation foundation began its operation.

If the general picture of senior citizens' clubs in Malaysia is predominantly about the Chinese ethnic grouping, older Malays and Muslin ethnic groups are entwined in a narrative that mirrors the development of religious schools, *madrasah*s and *pondok*s that are evolving with the times. A prominent case study would be the Complex for Older Persons under the al-Jenderami Foundation. The foundation itself was established in 2003, focusing on charitable and welfare pursuits, especially on Islamic and spiritual education of children and older persons. The al-Jenderami Foundation operates a facility for older women who spend their days in Quranic learning and prayer, and it is not an old folks' home or nursing centre. There are other such modern centres of Islamic learning for older persons, and they are very popular among the older Malays. The appeal of such religious learning facilities is in tandem with the common belief that the remaining time in old age is best spent in preparation for the life hereafter.

Prior to the latest general election in 2018 that resulted in a historical change in government, the previous National Front (BN) administration unveiled an initiative #mySeniors or Beautiful Life for Seniors as part of the National Blue Ocean Strategy to help older Malaysians lead a productive and meaningful life. One of the three segments is on Lifelong Learning for Seniors which aims to provide skills training to help older persons get age-friendly jobs or to enhance their lives. The Department of Social Welfare aims to expand the 59 Older Person Activity Centres (PAWEs) so that there will be one established in every parliamentary constituency as part of the

Beautiful Life for Seniors initiative. There is currently no change in the expansionary trajectory of PAWEs under the new government.

If the extensive concept of lifelong learning is visualised through a knowledge society model in which individuals of all ages are equipped and motivated to contribute in learning across diverse settings throughout the lifespan, we can conclude that the current situation in Malaysia remains under-developed. Lifelong learning for older persons focuses on self-care, self-improvement and self-fulfilment, but it must be reinforced in multiple ways in terms of the vision and philosophy, operational framework, funding structure and related national policy discourse (Quadagno, 2017). In principle, the policy on lifelong learning is related to human capital which enables individuals to be productive and continue in the employment longer, advance their skills, endeavour into a new career, as well as becoming more adaptive and creative towards life transitions and modifications. In the context of population ageing, a modern policy paradigm for lifelong learning goals is to promote physical, psychological and societal benefits including delayed cognitive decline; increased physical activity; improved life quality; as well as enhanced positive social relationships and conservation of social capital. Lifelong learning acquisition and active participation by older persons are evident towards a more positive experience of ageing in the society (Lunenfeld & Stratton, 2013).

In the first Pillar of the Madrid International Plan of Action on Ageing (United Nations, 2012), "Older Persons and Development", governments are called upon to provide opportunities, programmes and support to encourage older persons to participate or continue to participate in cultural, economic, political, social life and lifelong learning. Moreover, under the new Sustainable Development Goals (SDG), governments are required to ensure inclusive and equitable quality education and promote lifelong learning opportunities for all (Goal 4). There is no question that civic society has a responsibility and an obligation to ensure that older persons in Malaysia receive adequate resources and given opportunities to participate in lifelong learning activities.

The University of the Third Age in Malaysia

The origins and development of the University of the Third Age (U3A) programme in Malaysia have been published previously (Ibrahim, Hamid, Chai, Abdullah 2016; Ibrahim et al. 2017). In 2007, the Institute of Gerontology, Universiti Putra Malaysia piloted a Lifelong Learning Initiative for the Elderly (LLIFE) (Ibrahim & Hamid, 2012). In 2008, with support from the Government of Malaysia and United Nations Population Fund (UNFPA), the Institute set up the first U3A programme in the country. The U3A Malaysia programme is a hybrid of British and French models, utilising UPM's campus facilities and resources but promoting self-determination among participating members. An autonomous approach, as practised by the British model (Swindell & Thompon, 1995), was seen as a good way to help transfer the Institute's organisational responsibility in terms of finance, management and operations of current and future U3As. This was made possible through the empowerment of the U3A members in managing their own learning activities through registered not-for-profit organisations. In Malaysian U3As,

> ...there are no academic pre-requisites and anyone can join. Participants not only broaden their knowledge, skills and experience through the U3A programme, but also contribute and actively participate in their own learning and well-being. U3A Malaysia was envisioned right from the outset as a programme for senior citizens by senior citizens, and a membership structure was adopted so that a pro-tem committee could be established as early as possible. The lifelong learning programme aims to: (a) Optimise the potential, ability and experience of older Malaysians through lifelong learning; (b) Improve social, economic and cultural participation of older Malaysians towards an active and productive old age, and; (c) Increase the opportunities for older Malaysians to contribute towards national development.
>
> Ibrahim et al. (2016: 254)

At present, there are three registered U3As in Malaysia, namely U3A Kuala Lumpur and Selangor (2010), U3A Bandar Utama (2016) and U3A Petaling Jaya (2018). Of the three, only U3A Kuala Lumpur and Selangor is based at the UPM campus in Serdang while the remaining two are fully independent entities that rely on local community support. All the three U3As operate on a semester-based course structure, and annual certificate presentation ceremonies are held separately. U3A members pay their own fees to participate in weekly study programmes that last from four to six weeks each. On average, each course costs about RM50 (USD12), a significant increase from the original RM30 (USD7) per study programme in 2008 and participants bear their own material fees. All fees collected go to pay for instructors and rental of facilities, including the upkeep and management of operations. The Department of Social Welfare has been funding the U3A Kuala Lumpur and Selangor since 2013, but the amount represents a fraction of the cost involved. Nevertheless, the finances of all the U3As depends largely on the effort of the executive committee members who are elected through annual general meetings once every two to three years. Fundraising should be carried out to keep course fees low, or members will have to make up for the difference. This is not a desirable solution as higher fees will deter the participation of other older persons with lower income or savings.

Three key philosophies unite the U3As in Malaysia. First, no academic degrees are awarded or required for participation. Second, those who teach shall learn and those who learn shall teach. While many professional instructors are appointed, some members who are senior citizens themselves also teach a few study programmes. Lastly, U3A members decide what study programmes are to be introduced, continued or dropped. Each U3A operates by soliciting regular feedback from participating members and are held fully accountable through annual general meetings where the association's activities and finances are scrutinised. In comparison with its international counterparts, the U3As in Malaysia have a lower age limit as most accept members above the age of 50. For example, U3A Kuala Lumpur and Selangor's statute only accepts those aged 55 and above as ordinary members, while those between the age of 50 and 54 are recognised as associate members. Although the three existing U3As are not affiliated with one another now, the groups will have to eventually sit down to chart a path forward together as other lifelong learning associations are established nationwide.

Study programmes in the U3A programmes in Malaysia are varied and range from topics on health and exercises, languages and performing arts/music, arts, craft and living skills, agriculture, entrepreneurship, information technology as well as culture and religion. Study programmes were designed to include not only the components of leisure and interests that would attract older participants, but also to encourage the acquisition of new knowledge, skills and experience through the fusion of both academic and practical concerns. Teaching and learning strategies differ from one instructor to another, but most conform to traditional classroom activities with complementing outdoor excursions for some courses such as outdoor photography or urban sketching. One of the most common challenges in course planning and management is to balance the interests of new members and sustaining the need of existing, older members. The success of U3A's approach has led to calls for further replication and upscaling, but no specific federal funds were set aside or approved for such purpose.

Efforts to strengthen the advocacy role of U3As through community services and volunteerism are an on-going challenge although many members are active with various causes of their own choosing. The University of the Third Age is a platform for older persons to advocate on issues related to old age and ageing. At present, U3As in Malaysia are more focused on managing their own operations and activities for continued sustainability rather than involved in extension services. It must be noted that the U3A has become typical and is regarded as part of the goal for social inclusion of older individuals in many ageing societies globally, as what has been demonstrated by the UK, France, Germany, China, Australia and the Nordic countries. To echo the societal trend in ensuring the goals for knowledge, skills and leisurely pursuits, the lifelong learning policies and programmes manage the risks of an ageing population through a more inclusive educational system in later life. To date, the U3A programme has expanded the borders of knowledge, skills and experience among older individuals. The programme has encouraged the older Malaysians to be dynamic participants in their own learning process, but the associations need more time to mature and stabilize their operations before taking on broader roles.

The execution of lifelong learning policies for older individuals requires sector-wide approaches, which rely on a more structured funding for sustenance as an inter-ministerial or inter-agency programme impacting at the micro-level (Formosa, 2010, 2012a, 2016). The culture of dependency on the government for support has been long in existence in our sociopolitical systems. In addition, there is an unhealthy fixation on physical buildings for community activities rather than utilising existing resources. The drive to build or rent more facilities as Older Person Activity Centres stem from a misunderstood premise—namely that older persons lack public spaces for their activities. U3As or any senior centres should not feel hampered without clubhouse-like facilities but instead should focus on their members as an invaluable asset and partners in learning. Learning can happen anywhere, but building a facility will not guarantee sustained increased participation of older persons over time. At the institutional level, the idea of giving free handouts is viewed as short-lived as it is subjected to the changing personal fortunes of the patron or the relative importance of a particular political agenda. The third age learning movement must be prepared to discover new approaches to bridge the resources divide and devise more sustainable efforts.

Future of U3As in Malaysia

U3A constitutes a community-based lifelong learning programme that has a relevant role and place in Malaysia. However, this kind of activities requires support from government agencies and departments who can support the associations financially or in kind. Community colleges, for example, are a ready partner that can play a significant role in the upscaling of U3As. The government has outlined the need for replication of the University of the Third Age in the *11th Malaysia Plan* (2016–2020). Nevertheless, progress has been slow, and only three U3As have been established so far. This slow development may be attributed to the low pickup rate among the stakeholders. There is no proper funding or national structure that governs the development of the U3As. U3As in Malaysia are managed by an association of older persons solely established as a learning organisation. As a non-government organisation, the strength of a society to coordinate a learning programme depended on the ability of its members to manage itself and its activities. If members are only interested in signing up for study programmes and not the trouble to manage the programme, make it more accessible and leverage its position as an advocate for older persons, then its development will remain stunted. Indeed, in an earlier paper some of the authors herein have highlighted a number of macro- and micro-challenges for U3A movement in Malaysia:

> At the macro level, challenges persist such as societal transformation and cultural lag; awareness and participation of stakeholders; incorporation the changes and transformation within the formal education system; policy integration for lifelong learning into the wider education system; an institutional framework for improved access, coordination and monitoring; as well as adequate tangible and intangible incentives.... At the micro level, the U3A associations are now part of a structured, non-formal education in later life, alongside informal learning that is customary in the community. Both of these types of learning are not recognised in the formal education system and are relegated to be part of social welfare.
>
> Ibrahim et al. (2017: 54–5)

The Association for Lifelong Learning of Older Persons U3A Kuala Lumpur and Selangor has taken over the running of the programme from the Institute of Gerontology since 2012. Study programmes are mostly conducted within the Universiti Putra Malaysia's premise as well as other locations such as the Hulu Langat Community College and the National Arts and Heritage Academy. The U3A Bandar Utama established in 2016 was set up with their own by-laws and receives support from businesses and sociopolitical groups in the area. U3A Petaling Jaya was registered just recently in 2018 and has yet to organise its first annual general meeting. The challenges facing these U3A associations are more similar than different in many ways. Associations are stronger if they survive the personalities that drive them, and the measure of a successful organisation is in member mobilisation and succession planning. The Malaysian Research Institute of Ageing (formerly the Institute of Gerontology), UPM needs to play a coordinating role to promote the development of U3As nationwide and enlist the support and cooperation of relevant local stakeholders. There is a need to overcome replication issues by providing a platform to assist and promote lifelong learning among older Malaysians as a strategy to promote healthy, active and productive ageing. Coherence and understanding of the goals and purposes of U3As in Malaysia must be shared so that there is a broad buy-in of the initial, founding ideals. In the UK, a governing body called the National Third Age Trust was established which supports and is responsible after all U3As (Formosa, 2012b, 2014). The existence of this structure can help coordinate, monitor and support the development of the U3As in Malaysia. Sustained efforts and the role of key actors and champions, whether inside or outside the government, are an absolute necessity. Old modalities for replication to implement artificial or inflated figures must be discarded.

A variety of lifelong learning programmes are being offered by a number of providers such as government agencies, private sector organisations as well as non-government bodies. Policy-wise, the Alliance of Hope administration should seek to level the playing field and promote, as well as reward, lifelong learning activities that are sustainable, effective and receptive to the needs of older Malaysians throughout the country. Lifelong learning activities in Malaysia are still driven by policy supply rather than local demands from senior citizens living in the community. Older persons in care and nursing homes in Malaysia have little access to stimulating and invigorating daily activities. There is a clear need and demand for more learning activities for older persons, but few are dedicated to promoting and providing such services. The cultural and language differences among the multi-ethnic groups of older persons also reflected a need for varying approaches and strategies. Local governments, be it district offices or municipalities, must consider what it takes to develop truly age-friendly communities and cities and learn to promote sustainable lifelong learning initiatives for older persons.

Conclusion

The development of U3As in Malaysia is related to the focus of related government policies and programmes for older persons. Although highlighted as a model for replication, the U3A programme in Malaysia, specifically its modus operandi, is still something new as senior citizens face real difficulties in organising these activities for themselves. The challenges of upscaling U3A in Malaysia can be addressed through more training and documentation of past experiences and know-how on successfully running an organisation and lifelong learning programme at the same time. However, the real challenges on the ground have always been finances and manpower support.

The University of the Third Age programme in Malaysia renewed a focus on learning in later life and promoted a model where older persons themselves organise their own learning activities. As such, the three U3As or lifelong learning associations for older persons are not just running study programmes, but also managing the affairs of their own organisation. As the new lifelong learning associations in the country mature and stabilize, only time can tell if the U3A brand or model will survive and flourish in a new Malaysia. Perhaps the Universities of the Third Age in Malaysia will exist in many different forms than originally envisaged, or it will cease to replicate further without government support or fail to compete with other modalities in promoting active and productive ageing. Now that a seed has been planted, how the U3A tree will grow into a forest is up to the members and older persons themselves.

References

Department of Statistics Malaysia. (2015). *Malaysia economic statistics, 2015, time series*. Putrajaya: Department of Statistics Malaysia.

Department of Statistics Malaysia. (2016). *Population projections (revised), Malaysia 2010–2040*. Putrajaya: Department of Statistics Malaysia.

Department of Statistics Malaysia. (2017a, August). Population and demographics—Ageing. *BPPD Newsletter, 1*. Putrajaya: Population and Demography Division, Department of Statistics Malaysia.

Department of Statistics Malaysia. (2017b). *Abridged life tables, Malaysia, 2015–2017*. Putrajaya: Department of Statistics Malaysia.

Economic Planning Unit. (2005). *Ninth Malaysia plan: 2006–2010*. Economic Planning Unit: Putrajaya.

Economic Planning Unit. (2010). *Tenth Malaysia plan: 2011–2015*. Economic Planning Unit: Putrajaya.

Economic Planning Unit. (2015). *Eleventh Malaysia plan: 2016–2020*. Economic Planning Unit: Putrajaya.

Formosa, M. (2010). Universities of the Third Age: A rationale for transformative education in later life. *Journal of Transformative Education, 8*(3), 197–219.

Formosa, M. (2012a). European Union policy on older adult learning: A critical commentary. *Journal of Aging and Social Policy, 24*(4), 384–399.

Formosa, M. (2012b). Education and older adults at the University of the Third Age. *Educational Gerontology, 38*(1), 114–125.

Formosa, M. (2014). Four decades of Universities of the Third Age: Past, present, and future. *Ageing & Society, 34*(1), 42–66.

Formosa, M. (2016). Malta. In B. Findsen & M. Formosa (Eds.), *International perspectives on older adult education: Research, policies, practices* (pp. 161–272). Cham, Switzerland: Springer.

Hamid, T. A., & Chai, S. T. (2013). Meeting the needs of older Malaysians: Expansion, diversification & multi-sector collaboration. *Malaysian Journal of Economic Studies, 50*(2), 157–174.

Hamid, T. A., & Yahaya, N. (2008). National policy for the elderly in Malaysia: Achievement and challenges. In H. G. Lee (Ed.), *Ageing in Southeast and East Asia: Family, social protection and policy challenges* (pp. 108–133). ISEAS—Yusof Ishak Institute: Singapore.

Ibrahim, R., & Hamid, T. A. (2012). Experience of lifelong learning among participants of LLIFE programme. In T. A. Hamid, H. Sulaiman, & S. F. Z. Abdullah (Eds.), *Profile of older Malaysians: Current and future challenges* (pp. 280–303). Serdang: Universiti Putra Malaysia.

Ibrahim, R., Hamid, T. A., Chai, S. T., & Abdullah, S. F. Z. (2016). Malaysia. In B. Findsen & M. Formosa (Eds.), *International perspectives on older adult education: Research, policies and practice* (pp. 247–260). New York: Springer.

Ibrahim, R., Zakaria, N. S., Hamid, T. A., & Chai, S. T. (2017). On 'learning for leisure' and the margins of mainstream education: A critical review of the U3A movement in Malaysia. *International Journal on Ageing in Developing Countries, 2*(1), 41–58.

Lunenfeld, B., & Stratton, P. (2013). The clinical consequences of an ageing world and preventive strategies. *Best Practice and Research, 27*(5), 643–659.

Merriam, S. B., & Mohamad, M. (2000). How cultural values shape learning in older adulthood: The case of Malaysia. *Adult Education Quarterly, 51*(1), 45–63.

Ministry of Health. (2009). *Annual Report, 2009.* Ministry of Health: Putrajaya.

Ministry of Higher Education. (2011). *Blueprint on enculturation of lifelong learning for Malaysia, 2011–2020.* Putrajaya: Ministry of Higher Education.

Ministry of National Unity and Social Development. (1999). *National policy for the elderly.* Ministry of National Unity and Social Development: Kuala Lumpur.

Ministry of Women, Family & Community Development. (2011). *National policy for older persons.* Putrajaya: Ministry of Women, Family & Community Development.

Ong, F. S., Phillips, D. R., & Hamid, T. A. (2009). Ageing in Malaysia: Progress and prospects. In T. Fu & R. Hughes (Eds.), *Ageing in East Asia: Challenges and policies for the Twenty-First Century* (pp. 138–160). New York: Routledge.

Quadagno, J. (2017). *Ageing and the life course: An introduction to social gerontology* (7th Ed.). McGraw-Hill Education.

Swindell, R., & Thompon, J. (1995). An international perspective on the University of the Third Age. *Educational Gerontology, 21*(5), 429–447.

Tey, N. P., Siraj, S., Kamaruzzaman, S. B., Chin, A. V., Tan, M. P., Sinnappan, G. S., et al. (2016). Aging in multi-ethnic Malaysia. *The Gerontologist, 56*(4), 603–609.

United Nations. (2012). *Madrid international plan of action on ageing.* New York: United Nations.

World Bank. (2017). *World development indicators 2017.* Washington, DC: World Bank. https://openknowledge.worldbank.org/handle/10986/26447. Accessed 23 November 2018.

Tengku Aizan Hamid is Professor of Gerontology and founding Director of the Institute of Gerontology, Universiti Putra Malaysia, in 2002. In 2008, she initiated Malaysia's first U3A programme at Universiti Putra Malaysia's campus in Serdang with the support of UNFPA and the Government of Malaysia. Since then, she has been acting in an advisory capacity to U3A Kuala Lumpur and Selangor. In 2015, the Institute was renamed and designated by the Cabinet of the Malaysian Government as the 'Malaysian Research Institute on Ageing'. She is currently President of the Gerontological Association of Malaysia.

Noor Syamilah Zakaria is a Senior Lecturer at the Department of Counsellor Education and Counselling Psychology, Faculty of Educational Studies, Universiti Putra Malaysia. She is a Registered Counsellor and Certified Counselling Practitioner with the Board of Counsellors (Malaysia). She has been a Research Associate of MyAgeing since 2015, and her academic focus is foregrounded on counsellor education and supervision, with research interests in counselling ethics, teacher education, adult education and older adult lifelong learning.

Nur Aira Abd Rahim is Senior Lecturer at the Department of Professional Development and Continuing Education, Faculty of Educational Studies, Universiti Putra Malaysia. Her research interest and area of specialisation are in the areas of adult education and adult learning.

Sen Tyng Chai is Senior Research Officer at the Laboratory of Social Gerontology, Malaysian Research Institute on Ageing, Universiti Putra Malaysia. He has served as Project Officer for the UNFPA-funded activity on 'Lifelong Learning for Older Malaysians' from 2008 to 2012.

Siti Aisyah Nor Akahbar is Science Officer at the Laboratory of Medical Gerontology, Malaysian Research Institute on Ageing, Universiti Putra Malaysia.

Chapter 17
Universities of the Third Age in Aotearoa New Zealand

Brian Findsen

Introduction

To better understand the developments of the University of the Third Age (U3A) movement in Aotearoa New Zealand, a bi-cultural (and arguably, multi-cultural) country, it is necessary to consider the characteristics of society more generally, including the place of older people in a European-colonized country. As for any agency, the U3As have emerged from particular contextual circumstances of history, geography, demographics and the political economy. What it means to be "old" in this society varies across ethnicities (e.g. indigenous Māori, Europeans, Pacific peoples, Asians), and the ways in which many older people negotiate their daily lives are linked closely to active social networks such as the U3As. This chapter explores the development of the U3A movement in that wider context before looking into what constitutes older adult learning/education. While there are unique organizations, such as the U3A, that prioritize senior citizens as their primary constituency, there are a host of other adult education agencies whose work and programmes may have relevance too. Amid the middle class of New Zealanders, the majority of whom are Pākehā (European extraction), there is active contestation for participants for lifelong education. After surveying the array of providers displaying varying philosophical orientations and associated practices, this chapter focusses on the character of U3As, especially in terms of pertinent historical conditions, current structures and practices. In the final section, the chapter investigates some current issues facing the U3A movement in this country and how U3As contribute to older adulthood in this society.

B. Findsen (✉)
Division of Education, University of Waikato, Hamilton, New Zealand
e-mail: bfindsen@waikato.ac.nz

© Springer Nature Switzerland AG 2019
M. Formosa (ed.), *The University of the Third Age and Active Ageing*, International Perspectives on Aging 23, https://doi.org/10.1007/978-3-030-21515-6_17

Society in Aotearoa New Zealand

No credible history of this country can neglect the fundamental relationship between *tangata-whenua*—"people of the land" (indigenous Māori)—and the colonizer, the British. While the indigenous people have been here in Aotearoa for centuries, most of the recent settlement by Europeans has been concentrated from the early 1800s (King, 2004). Suffice to acknowledge, it is a history fraught by contestation for land, epitomized by wars in the nineteenth century, mainly between the colonizers and colonized. The Treaty of Waitangi signed in 1840 signalled a new relationship based on partnership, protection and participation (principles now embedded in educational policy today). Many of the current social and educational institutions in this country are of British origin (Dakin, 1992), yet Māori have developed their own lifelong learning projectory through the establishment of *kōhanga reo* (language nests), *kura kaupapa Māori* schools and *whare wānānga* (tertiary education institutions). This parallel development has been especially strong since the 1980s when a renaissance occurred to boost self-determination in which *te reo* (Māori language) has enacted an important function. This bi-cultural history is relevant in terms of comprehending why Māori have not embraced the U3As, identified as agencies of the oppressors (Freire, 1972).

The above brief description of Pākehā and Māori relations, essentially a bi-cultural relationship, obscures another reality. In the last three-four decades, New Zealand has become a multi-cultural society, with Asian migration into the major cities as a prominent manifestation of increased diversity (Boston & Davey, 2006). In the most recent census, the respective percentages of ethnic groups were as follows: European 74%; Māori 14.9%; Pacific Peoples 7.4%; Asian 11.8%; Middle Eastern, Latin American, African 1.2% and other 1.7% (Statistics New Zealand, 2013). These statistics reveal the increasing diversity of the New Zealand population, including that of seniors. Alongside this changing ethnicity, corresponding ageing structures also have occurred. In 2013, people over the age of 65 were at 14.3% compared to 12.1% in 2001. The very recent statistics for the 2018 census are not yet available, but indicators point to proportionately more people over the age of 65, together with an increasing life expectancy.

While defining "older people" or "seniors" or "elders" is always problematic (Phillipson, 1998, 2013), in the New Zealand context the award of a universal pension at age 65 is a crude indicator for retirement. Compulsory retirement has been outlawed and given the neo-liberal political and economic environment in which many people beyond 65 find themselves, more people are staying on in jobs, seeking encore careers or looking for part-time work to try to secure their economic futures (Beatty & Visser, 2005; Boston & Davey, 2006). Older people are living in the "risk society" (Beck, 1992) where uncertainty is prevalent. The current New Zealand Government (formed in 2017 by a coalition of Labour, the Greens and New Zealand First parties, and led by a young woman) is conscious of rapidly changing circumstances and is actively seeking input into the development of a New Zealand Positive Ageing Strategy (Tobias, 2017). Familiar rhetoric of active ageing and of the heightened well-being of seniors is promoted by government; this is yet to be matched by much-needed resources in health, social welfare, housing and education (from early childhood to tertiary levels).

As for many other countries, the advent of neo-liberalism from the late 1980s onwards has heralded renewed focus on the minimal role of the state demonstrated through deregulation, heightened accountability, increased surveillance and a cult of efficiency (Olssen, Codd & O'Neill, 2004). The impact of these changes has not been experienced uniformly—the more marginalized members of society such as Māori, Pasifika, state beneficiaries and significant numbers of older people tend to experience the brunt of economic policy and practices. For many older people in 2018, the harsh reality is that housing and the cost of living have become prohibitively expensive (especially in larger urban areas where most live) and discretionary income is minimal. Hence, after nine years of a National-led government in which neo-liberalism has been promoted, the gaps between rich and poor have grown wider (Kelsey, 2015). The older generations are not exempt from these economic pressures.

(Older) Adult Education

Forms of adult education in Aotearoa New Zealand are truly diverse. Some of the providers of adult education, such as the U3As, have older people (retirees) as their primary focus as participants. Other examples include Seniornet and some 60 plus continuing education groups which may have spun away from their allegiance to a university (and are not considered as U3As). The men's shed movement has grown in popularity and helps meet the needs of some older men and unemployed younger men (Golding, 2015). Some of these informally established groups share some of the characteristics of U3As such as friendship, low costs and curricula autonomy (Findsen, 2005, 2006). Other adult education programmes offered by generalist education agencies may have programmes that appeal to the older persons' market. Such organizations, now less prevalent in New Zealand since the previous government trimmed funding to them by 80% around 2012, are high schools that have adult and community (ACE) education services. These providers have always appealed to local seniors because of their close proximity to home and relevance of courses to daily life (e.g. Coping with Grief; Eat Well for Less). Yet another category of provider includes agencies which have a social mission in which education/training has a prominent function. For instance, the Age Concern organization, in some of its 34 national locations, offers lifelong learning programmes for older citizens. As I have observed elsewhere, "some education is provided *with* and *by* older adults; some is provided *for* older adults" (Findsen, 2016: 300). This distinction of the source of the education is significant when the U3A movement in this country is analysed below.

Given the history and development of New Zealand society described earlier in this chapter, the preponderance of institutions involved in any way (directly or indirectly) is of British/European origin, as more fully explained by Dakin (1992). The *derivative* forms of adult education in this country such as the University Extension departments in universities (more latterly converted into centres for continuing education, now mostly defunct) echo the original mission of the source but have been subject to "Kiwi" modification. In many ways, the U3A movement in Aotearoa New Zealand follows this pattern of adaptation and might be characterized by the objects set out originally by the British pioneers such as Peter Laslett, Eric Midwinter and Michael Young (Midwinter, 2004). The U3As fit into the "model" of the adult and community education humanistic forms of provision promoted by ACE Aotearoa (the national body for adult and community education).

There is a necessity to provide balance in this explanation of adult education for seniors with regard to the indigenous people. Traditional Māori knowledge has always incorporated lifelong and life-wide dimensions, expressed in the term, *ako* (to learn and to teach). In a bid to sustain their distinctive cultural values and aspirations through *tino rangtiratanga* (self-determination), Māori have established a credible parallel system of education from early childhood through to late adulthood as explained above. In terms of the continuity of Māori knowledge, the *marae* (communal meeting place) occupies a central position wherein kaumātua (seniors) occupy a significant role as disseminators of local and ancestral *kaupapa* (purposes/philosophy) (Findsen & Tamarua, 2007).

A manifestation of a Māori seniors' learning space is the Hamilton-based Rauawaawa Trust. The Trust is an exemplar of an urban holistic entity where the needs of local kaumātua are of primary concern. As for the parallel Pākehā Age Concern Hamilton organization, this Trust offers a range of services to seniors to include education, social welfare, housing and hospitality. In educational terms, the programme was originally supported by the University of Waikato through its Centre for Continuing Education in a promotional partnership where funding came via the University to the Trust and much of the learning activity was controlled by Māori for Māori. (See, Findsen, 2014, for a fuller explanation of this education programme for Māori seniors). The purpose of this explanation is to point out that Māori people have taken ownership of much of their learning with and for kaumātua (Māori elders) and would see little relevance of U3A for meeting their cultural needs.

As will be understood when explaining how the U3A movement works in this country, their existence is primarily located in the *expressive* forms of education (Jarvis, 2001; Laslett, 1991). Arguably, U3As are quintessentially adult liberal education agencies founded on principles of individuals' considerable freedom in later life to pursue interests (perhaps long-suppressed, perhaps newly emergent) devoid of many of the complications of the second age. The third age offers a time "when adequate resources, adequate health, and few responsibilities provide a context for self-fulfilment, freedom, and purposeful engagement" (Rubinstein, 2002, p. 29). Yet society has changed considerably since the days of Peter Laslett and his ideas, expressed in *A fresh map of life* (1991). Increasingly, in Aotearoa New Zealand, later life patterns of living (indeed *surviving* for those in the lower socio-economic realms) have diversified as life expectancy has increased. Many older people (defined arbitrarily as beyond 65—the age of universal superannuation in New Zealand) out of necessity or choice decide to stay in the workforce to sustain a decent standard of existence. While some older workers may face particular issues in the workforce, such as ageism and on-going demands of ICT developments, there is plenty of learning that occurs through professional development and informal mentoring in workplaces (Beatty & Visser, 2005). For some seniors, part-time work options may be available, thus freeing them to pursue other pursuits such as learning for leisure and recreation (Elsey, 1986), epitomized by engagement in the U3A movement.

The Development of the U3A Movement

Philosophical Antecedents

In relation to the often-expressed themes of lifelong learning (economic sufficiency; self-fulfilment; active citizenship and social inclusion—see Findsen & Formosa, 2011), there is little doubt that the liberal adult education goal of individual self-fulfilment is paramount for the U3A movement. In the principles espoused in the UK and transported to this country "down-under", I argue that the philosophical base of the U3A movement has held steady with some local modifications. The five guiding principles referred to by Midwinter (2004) are quite expansive in character and are suggestive of a broad concern for the relationship of ageing and learning. A somewhat abridged and paraphrased version of these principles is as follows:

- *The University of the Third Age consists of people who learn and help others to learn. Those who teach are encouraged to learn and those who learn are encouraged to teach.* In effect, this principle highlights the peer learning context and equality of status of learners and teachers. It echoes the concept from Freire (1972) of the student–teacher and the teacher–student.
- *Joining a group is by personal choice, and no qualification is required to enter. Standards are maintained internally and the form of education decided by the individuals in that group.* This principle emphasizes the autonomy of each group to decide its curriculum and reinforces the equality among members.
- *A minimal fee is expected of everyone of "the University", and while fundraising is expected of members, no approach is to be made to local or central government.* This principle again treasures the independence of the group, not wanting any political interference that may derive from accepting funds from government.
- *No payment is to be made for teaching; "arrangements will be maintained"* (Midwinter, 2004: 47) *with other adult education providers of programmes.* This principle highlights the voluntary contribution of individuals; it also pays heed to the work of other adult education agencies in achieving complementary goals.
- *The curriculum shall be as wide as "human and financial resources permit"* (Midwinter, 2004: 47); *strong emphasis will be placed on research projects, practical skills and leisure pursuits; insistence on learning as an end in itself.* This principle acknowledges the breadth of curriculum through emphasizing the worth of learning for its own sake. The emphasis on "research projects" is commented upon later.

In the New Zealand context, the original intent of the U3A is upheld, as exemplified by the front page of the U3A New Zealand (U3A NZ) website. Principle 1 above is virtually repeated word for word. It is explained that the "U" in U3A means a University defined as "a group of people who gather together to study a particular topic of interest" but not a university in a legislated sense. From the U3A NZ website, the purposes of U3A are to:

- Encourage further learning by listening, understanding, contributing, researching and participating in discussion and relevant excursions about new topics,
- Tap the great reservoir of knowledge, skills and experience of retired men and women and
- Provide a venue for the meeting together of like-minded people to learn, contribute and to make new friends in their community.

(i)   U3A NZ, (n.d.)

Elsewhere on the U3A NZ website, the concept of "third age" is explored in a fairly rudimentary manner. The first age is described as "childhood", the second as "career" and the third as "retirement". These demarcations tend to reinforce the notion of an ideal (arguably, outmoded) age differentiated age structure of young (education), middle (work) and old (leisure), explained by Riley and Riley (1994). These authors point to a newer "age integrated" model of parallel consideration of education, work and leisure at any point of life. So, in some respects, the U3A draws upon the older conceptualization of older age as a separate, distinctive feature of the life course, in which "retired" people join together for learning and socializing.

The Characteristics of U3As

In New Zealand, there are 82 U3As across the entire country, 26 of which are in Auckland. The proportion of Auckland U3As compared with the rest of the country reflects the population base. Around one and half million people live in the metropolitan Auckland area and the entire population of the country is around 4.8 million. The first U3A in New Zealand was established in Remuera, Auckland, a prosperous suburb in the inner city in 1989. It is explained that the New Zealand model is most closely aligned to the English, where "this model promotes increased personal learning, confidence and enjoyment by reading, research, discussions and field trips" (NZ U3A, n.d.). In reality, the vast majority of U3As are located in cities or towns sufficiently large to make a U3A viable.

What Do U3As Do?

As explained on the main NZ U3A (n.d.) website, the main pattern of interaction between members consists of a monthly meeting (usually comprising of an invited guest speaker of national/international standing) followed by presentation(s) by members and meetings of special interest groups (SIGs). The position of special interest groups is important as they provide opportunity for individuals to focus on topics of salience to them while not being in areas of interest to the entire community. According to the official website, these SIGS meet regularly (once or twice a month), have between 6 and 20 members (venue size permitting), the programme is decided by members and each member is expected to be an "active learner". Examples of SIGs are art for fun, inventors, archaeology, medical science, modern history, literature, films, ancient history, travel and others. As clearly enunciated, these SIGs are "U3A's point of difference from other groups for seniors". At the Browns Bay (Auckland) U3A, it is proudly proclaimed that there are 26 SIGs available. As a seaside suburb located north of the Auckland CBD in a fairly robust socio-economic area, the extent of U3A membership is high, thus reflecting the principal recruits from within the nation—white, middle-class, primarily professional, retired people.

The management of U3As is sufficient to keep the organization running wherein a committee is elected annually consisting of up to 10 members with varied responsibilities such as looking for new members, conducting a newsletter, catering, publicity, organizing speakers and maintaining a website. It is clear that some tasks are shared across local U3As and between the national and the local. It is interesting to observe, given comments made earlier in this chapter about the character of older adult education in this country, that the website acknowledges this diversity of older adult education providers. Other groups which offer membership to "retired folk" consist of continuing education classes, service groups such as Rotary and Lions, sports clubs (even broadening to chess and crafts) and various social groups. Any member can belong to a SIG and can join numerous groups, if having sufficient time. It is pointed out that "real socialising takes place in these small groups", many meetings held in members' homes or community centres. Importantly, "by doing this we strive to fulfil the spirit of U3A as stated by Peter Laslett who founded U3A" (U3A NZ, n.d.).

One of the impressive features of the U3A movement in New Zealand has been the publication of the *U3A Information and Starter Kit* (May 2015) which sets out the history, objectives and principles of the U3A, the nature of study groups (SIGs), establishment and recruitment, the work of steering committees, a constitution and even some adult learning principles. These principles appear quite derivative of Knowles (1980) to emphasize the importance of informality, a supportive emotional environment, diversity of learning modes and drawing upon members' previous life experiences. An important caveat is that "the quality of learning is not so much dependent upon the quality of the information provided but upon the quality of the relationships that exist between people in the group" (U3A, n.d.: 7).

Issues Facing U3As in Aotearoa New Zealand

A commonplace but nevertheless pertinent critique of the U3A movement is that it has been captured by the middle class (Formosa, 2000, 2014, 2016; Findsen, 2002;

Formosa & Higgs, 2015). Bourdieu (1974) in his explanation of the character of cultural capital illustrates how older adults who are not part of the U3A mainstream (in this case, Pākehā middle-class, primarily women) are likely to feel psychological discomfort and cultural dysfunction in such learning contexts. While there is no overt barrier to participation by minorities (e.g. Pasifika, Māori and Asian), the reality is that the vast majority of U3A adherents are from the dominant cultural group. There does not appear to be any overt discriminatory practice on behalf of U3A NZ but nor is there active recruitment of say, Pasifika, in languages aside from the dominant English. The description of the Rauawawawa Trust (see above) involving Māori elders reflects the strong preference by minorities for same ethnicity networks and peers among older learners. Hence, the prevalence of Pākehā participants is likely to continue. In addition, while the curriculum is highly feminized, it is likely that women will continue to form the majority of members (Withnall, 2010).

The autonomy of the U3A movement in New Zealand is especially strong. Meaningful connections with neighbouring universities or other tertiary education institutes are rare and judged unnecessary. U3As are adept at self-governance and self-management. The deliberate low cost and fluidity of membership attract people who may not wish to have strong ties but nevertheless enjoy learning alongside others with similar interests. The voluntary ethic undergirding U3A NZ is very much a strength and helps to support older adult education even if judged from an economic proposition. Swindell (2000) estimated that in terms of actual monetary value way back at the turn of the millennium, the volunteer contribution (based on a notional $10 per hour) was easily over half a million dollars. That figure would be several million dollars now. As he rightly pointed out, the knowledge resource of retired people bringing their expertise and knowledge to this non-formal education agency has meant that paying "second age" teachers was unnecessary (ibid.: 24). However, the flip side of voluntarism is reliance on uncertain economic support and detachment from government funding sources, slim and contestable as they may be.

In the original constitution of the U3A movement based on the English model, the expectation that research would be carried out was strong. Given Laslett's (1991) own context of a university when he helped to begin the movement, it is not surprising that research figured prominently. In the New Zealand context, a location where pioneering practicality has been highly valued, research with a capital "R" has never been to the fore. However, given the increasing SIGs in the urban areas, there is likely to be some less formalized research undertaken to support the development of specific subjects. A downside to more distanced relationships with a university is the comparative lack of synergy often triggered by mutual research interests of academics and volunteer practitioners. Another latent prospect for the U3A movement is its promotion of public policy affecting the provision of older adult education. Given the overall potency of the movement as a provider of elder education in this country, there is potential for influencing the new government in both the Positive Ageing Policy development—there is currently a new Positive Ageing Strategy being developed by the Coalition Government—and in the lifelong learning space. There is a faint connection only with the national organization for adult learning in this country, ACE Aotearoa. On both fronts—active ageing and lifelong learning—the U3As are in a unique position to exert political influence without giving away their treasured independence.

Contributions of U3As

Swindell (2005: 33) asserted in respect to Australian U3As that they "can surely lay claim to being the "Jewel in the Crown" among adult education success stories for older learners" and that their "do-it-yourself philosophy" has boosted their immunity from major problems often faced by other providers such as funding and challenges to their autonomy. As he argued in a more recent paper,

> For more than 20 years, with little or no systematic support from government or funding agencies, U3As in Australia and NZ have quietly provided many, very-low-cost opportunities for members to take part in most or all of the successful ageing activities that are associated with continued independence in later life. There are now more than 270 U3As in Australasia and the number of independent groups and total membership continues to increase. Successive ageing cohorts are better educated than earlier cohorts and it seems reasonable to speculate that future retirees will continue to be attracted to meaningful activities involving the latter part of the lifelong learning continuum, because of the many benefits associated with keeping the mind active and mixing with lively, like-minded colleagues. Few organisations for retirees can point to a similar range of mentally, physically and socially stimulating courses, and the wide variety of volunteering opportunities provided by most U3As.
>
> Swindell, Vassella, Morgan, and Sayer (2011: 200)

Swindell's sentiment also pertains to the U3A movement in Aotearoa New Zealand. Clearly, U3A NZ is the leading agency for older adult education, but just as importantly, it is not the only source of education for elders. In respect to the expressive form of adult education, U3As provide a solid platform for liberal adult education for elders. The fundamentals of autonomy, flexibility and creativity are at the heart of what they are achieving. These agencies continue to provide active learning contexts for third agers in which cognitive, social and emotional aspects (Illeris, 2004) are all incorporated holistically into the learning environment. Importantly, too, the consolidation of social capital in later life is not to be under-estimated (Field, 2003), especially as social isolation among older people is currently identified by Age Concern New Zealand as a primary challenge for New Zealand society to confront. Those in the third age will eventually enter the fourth age and the relationships strengthened while comparatively younger in the U3A may be vital as dependence on others returns in much older age. Nevertheless, it remains to be seen whether the economic realities in which larger numbers of older people continue to work beyond 65 and who engage in a more instrumental pattern of learning will have any noticeable impact on the nature of U3A curriculum and associated pedagogy. Will what Elsey (1986) called the "leisure-recreation" model of adult education, which U3A epitomizes, be seriously challenged? Will the advent of competent third age people adept at ICT and distance learning have an effect on what is mainly a face-to-face form of peer learning? Will the changing composition of elders due to immigration patterns, a more diversified society and increased longevity have significant consequences for the work of U3As in Aotearoa New Zealand? These questions remain to be answered.

References

Beatty, P. T., & Visser, R. M. S. (Eds.). (2005). *Thriving on an aging workforce: Strategies for organizational and systematic change*. Malabar, FL: Krieger.

Beck, U. (1992). *Risk society: Towards a new modernity*. London: Sage.

Boston, J., & Davey, J. A. (2006). *Implications of population ageing: Opportunities and risks*. Wellington: Institute of Policy Studies, Victoria University of Wellington.

Bourdieu, P. (1974). The school as a conservative force. In J. Eggleston (Ed.), *Contemporary research in the sociology of education* (pp. 32–46). London: Metheuen.

Dakin, J. (1992). Derivative and innovative forms of adult education in Aotearoa New Zealand. *New Zealand Journal of Adult Learning, 20*(2), 29–49.

Elsey, B. (1986). *Social theory perspectives on adult education*. Nottingham: Department of Adult Education, University of Nottingham.

Field, J. (2003). *Social capital*. London: Routledge.

Findsen, B. (2002). Developing a conceptual framework for understanding older adults and learning. *New Zealand Journal of Adult Learning, 30*(2), 34–52.

Findsen, B. (2005). *Learning later*. Malabar, FL: Krieger.

Findsen, B. (2006). Social institutions as sites of learning for older adults: Differential opportunities. *Journal of Transformative Education, 4*(1), 65–81.

Findsen, B. (2014). Older adult education in a New Zealand university: Developments and issues. *International Journal of Education and Ageing, 3*(3), 211–224.

Findsen, B. (2016). New Zealand. In B., Findsen, & M., Formosa (Eds.), *International perspectives on older adult education: research, policy and practice* (pp. 297–308). Cham, Switzerland: Springer.

Findsen, B., & Formosa, M. (2011). *Lifelong learning in later life: A handbook on older adult learning*. Rotterdam: Sense Publishers.

Findsen, B. & Tamarua, L. (2007). Māori concepts of learning and knowledge. In S. B., Merriam & Associates (Eds.), *Non-Western perspectives on learning and knowledge* (pp. 75–97). Malabar, FL: Krieger Publishing Co.

Formosa, M. (2000). Older adult education in a Maltese University of the Third Age: A critical perspective. *Education and Ageing, 15*(3), 315–339.

Formosa, M. (2014). Four decades of Universities of the Third Age: Past, present, and future. *Ageing & Society, 34*(1), 42–66.

Formosa, M. (2016). Malta. In B. Findsen & M. Formosa (Eds.), *International perspectives on older adult education: Research, policies, practices* (pp. 161–272). Cham, Switzerland: Springer.

Formosa, M., & Higgs, P. (Eds.). (2015). *Social class in later life: Power, identity and lifestyle*. Bristol: The Policy Press.

Freire, P. (1972). *Pedagogy of the oppressed*. Hammondsworth: Penguin.

Golding, B. (2015). *The men's shed movement: The company of men*. Champaign: Common Ground Publishing.

Illeris, K. (2004). *The three dimensions of learning*. Denmark: Roskilde University Press.

Jarvis, P. (2001). *Learning in later life: An introduction for educators and carers*. London.

Kelsey, J. (2015). *The fire economy: New Zealand's reckoning*. Wellington: Bridget Williams Books with the New Zealand Law Foundation.

King, M. (2004). *The Penguin history of New Zealand*. Auckland: Penguin Books.

Knowles, M. S. (1980). *The modern practice of adult education: From pedagogy to andragogy* (Revised ed.). New York: Associated Press.

Laslett, P. (1991). *A fresh map of life: The emergence of the third age* (Revised ed.). Cambridge, MA: Harvard University.

Midwinter, E. (2004). *500 beacons: The U3A story.* London: Third Age Press.

Olssen, M., Codd, J., & O'Neill, A. (2004). *Education policy: Globalisation, citizenship and democracy.* London: Sage.

Phillipson, C. (2013). *Ageing.* Cambridge, United Kingdom: Polity Press.

Phillipson, C. (1998). *Reconstructing old age: New agendas in social theory and practice.* London: Sage.

Riley, M. & Riley, M. (1994). Structural lag: Past and future. In M. W. Riley, R. L Kahn, & A. Foner, (Eds.), *Age and structural lag* (pp. 15–36). New York: Wiley.

Rubinstein, R. L. (2002). The third age. In R. S. Weiss, & S. A., Bass (Eds.), *Challenges of the third age: Meaning and purpose in later life* (pp. 29–40). Oxford: Oxford University Press.

Statistics New Zealand (2013). *Stats NZ Tatauranga Aotearoa.* https://www.stats.govt.nz/. Accessed 20 August 2018.

Swindell, R. (2000). Half a million dollars a year: Voluntarism within Universities of the Third Age. *The New Zealand Journal of Adult Learning, 33*(2), 23–49.

Swindell, R. (2005). Some new directions in adult education programs for older Australians. *The New Zealand Journal of Adult Learning, 33*(2), 26–46.

Swindell, R., Vassella, K., Morgan, L., & Sayer, T. (2011). University of the Third Age in Australia and New Zealand: Capitalising on the cognitive resources of older volunteers. *Australian Journal on Ageing, 40*(4), 196–201.

Tobias, R. (2017). Fifty years of learning by older adults in Aotearoa New Zealand. *Australian Journal of Adult Learning, 57*(3), 440–457.

U3A NZ. (n.d.). *What is U3A?* https://www.u3a.nz/what.htm. Accessed 20 August 2018.

Withnall, A. (2010). *Improving learning in later life.* London: Routledge.

Brian Findsen is Professor of (Adult) Education, Division of Education, University of Waikato, New Zealand. He has worked in the field of adult and continuing education for over 35 years, primarily in his home country of New Zealand but also in Glasgow, Scotland, from 2004–2008. His main research interests include learning in later life, the sociology of adult education, social equity issues and international adult education. He has (co)written and edited several books, been guest editor for journals of special issues and additionally published extensively in later life learning.

Chapter 18
Third Age Education and the Senior University Movement in South Korea

Soo-Koung Jun and Karen Evans

Introduction

The growing attentiveness to opportunities for lifelong education for the older population reflects the demographic trends in the developed world and much of the developing world. According to the standard definition of the United Nations, societies in which the proportion of the population aged 65 and over is greater than 7%, 14% or 20% can be, respectively, characterised as 'ageing societies', 'aged societies' and 'super-aged societies'. South Korea qualified as an 'ageing society' in 1999 and is thus expected to become an 'aged society' before 2020 (Korean National Statistical Office, 2005). By comparison, the UK became an 'ageing society' in 1929, an 'aged society' in 1976 (preceding Korea by a total of forty-two years) and is projected to become a 'super-aged society' by 2021 (ibid.). While a few decades ago South Korea had the youngest population among OECD countries, the country is undergoing a rapid ageing process and is likely to be a country with a rapid demographic transition from an 'ageing' to an 'aged' society. This phenomenon—which Byun (2004) termed as 'intensive ageing'—does not provide enough time for a society to prepare for the demographic and social changes that will occur. Effective government and community-based action are critical to a proactive response to the situation. As a result, in 1981, the Korean Older People's Association (KOPA) established the Senior University. This chapter outlines the origins, characteristics and structures of the Senior University and the perspectives that have guided its development. The objective to promote active ageing in the population makes a strong case for government to continue to support Senior Universities and, in doing so, to prioritise the least advantaged group of South Korean citizens. It is argued that there is also a

S.-K. Jun
Namseoul University, Cheonan, South Korea

K. Evans (✉)
UCL Institute of Education, London, UK
e-mail: karen.evans@ucl.ac.uk

© Springer Nature Switzerland AG 2019
M. Formosa (ed.), *The University of the Third Age and Active Ageing*, International Perspectives on Aging 23, https://doi.org/10.1007/978-3-030-21515-6_18

role for the encouragement of self-help and mutual aid among those who are able to organise with less support and less intervention from the government. This chapter draws on other comparative research (Jun, 2014; Jun & Evans, 2014) in which the authors explored the scope of third age 'universities' in contributing to the support of older adults in such emergent scenarios.

Policy on Third Age Learning in South Korea

In South Korea, two governmental departments are in charge of older adult education: the Ministry of Education (MoE) and the Ministry of Health and Welfare (MoHW), also responsible for the 'Lifelong Education Law' (LEL) and 'Elderly Welfare Law' (EWL), respectively. On the one hand, education for older people is under the auspices of the LEL, as the field is considered to be a sub-field of lifelong education, though it is noteworthy that there is no direct mention of third age learning in the law. The LEL was preceded by the Social Education Law, which was introduced in South Korea in 1982, and subsequently embedded in the LEL. The latter was enacted in 1999 and revised in 2007. Article 2 Clause 1 of the LEL defines 'lifelong education' as 'all types of systemic educational activities other than regular school education', thus including diplomas, basic adult literacy education, vocational capacity-building education, liberal arts education, culture and arts education and education in civic participation. In terms of policy implementation, the 2007 LEL revision presents an administrative structure that works across central government, metropolitan governments and municipal governments. The National Institute for Lifelong Education (NIFLE) was launched in 2008 at central government level, although its name was changed into the National Institute for Lifelong Education (NILE) in 2012. Under the LEL, the NILE has been given full responsibility for promoting lifelong education in South Korea. Some of the NILE's primary responsibilities included, in 2016, the establishment of comprehensive national Master's programmes for the promotion of lifelong learning; development and support for lifelong education programmes; training of lifelong learning educators and personnel; building of networks between lifelong learning organisations; and support for metropolitan/provincial institutes for lifelong education. Under the supervision of metropolitan and provincial institutes for lifelong learning, the number of lifelong learning halls was 444 in 2016. On the other hand, the EWL was introduced in 1981. EWL institutions related to third age learning are classified as senior leisure facilities. Article 36 in the EWL classifies three institutions under seniors' leisure facilities: the senior community hall, the senior class and the senior welfare centre. In 2016, the number of senior classes researched by MoHW was 1393; the number of senior community halls was 65,044; and for senior welfare centres, the number was 350   (http://www.index.go.kr). The specific regulation of senior classes, as per the EWL, is shown in Table 18.1.

The senior welfare centres provide various service packages including learning programmes for seniors. The senior class is a form of senior-learning institution defined in the EWL as 'a facility intended to provide learning programmes related

Table 18.1 Provisions of Elderly Welfare Law for senior classes

Feature	Stipulations in Elderly Welfare Law (EWL): Articles 26 and 36
Intended users	Older adults aged over 60 years of age
Frequency	Classes/programmed activities provided at least once per week
Facilities	Space sufficient for more than 50 students Lecture room with office and lounge room/shared space Restroom
Staffing	Facility director and lecturer

Sources Based on Jung, Kim, and Kim (2014, p. 113)

to healthy hobbies, health maintenance, income security, and daily life of seniors, to meet older people's social participation needs'.

Historical Formations that Have Influenced the Development of Senior Universities

Generally, the terms 'Senior Universities', 'senior schools' and 'senior class' have the same meaning in South Korea—differentiated primarily by size—with the Elderly Welfare Law classifying all these groups as 'senior classes'. The first educational programme for older people was established in 'Hanul Senior University' in Bumil-dong (Busan) in 1970, and in 1972, a senior class was launched in Seoul Pyungsang Goyuck Won in Taewhagwan (Jongro, Seoul) (Han, 2015). The class, which was held once a week, was so successful that it attracted much attention and participation. Eventually, in 1973 a similar initiative was organised, entitled 'Duckmyungeusuk' in Catholic Yeosunghowegwan in Myungdong (Seoul). In 1972, Dr. Ha Doo-Chul set up 'In-Wang Senior University' at Seoul Nursing College in Seodaemoon-gu (Seoul), but later changed its name to the Korean Adult Education Association (KAEA) to establish senior schools on a nationwide scale, with the assistance of the MoE. KOPA was established in 1969 to establish and promote senior welfare services. In 1975, Park Jae-Gwan started working for this group and under his direction KOPA branches began setting up Senior Universities on a nationwide scale. In 1978, the MoE intensified its effort to provide older education nationwide, allowing for 7371 elementary schools to provide older education programmes in their community, setting up senior classes for liberal education to promote both community involvement and vocational education (KOPA, 1989). KOPA took over senior classes with government financing from the MoE and arranged the merger of the two national organisations for older people, the KAEA and the Korean Senior School Association (KSSA), into KOPA. In 1983, the KAEA had 120 Senior Universities (Park, 1999).

The intention was for the MoE senior classes to cater for younger and older people in equal measures. According to the government official who organised the senior classes programme, the classes were initially planned to teach filial piety to young people and secondly to provide new facilities similar to 'kyungrodang'—the traditional leisure places for older people. As a result, each elementary school had a 'senior classroom' and the room was equipped with *Baduk* (Korean checkers), *Janggi* (Korean chess), televisions, books and other effects. The planners envisaged a scenario of inter-generational learning where younger learners are taught culture and traditions by the senior learners. Between 1960 and 1980, when South Korea fell under a military regime, the government prohibited non-formal or informal educational organisations from offering social or liberal education. All learning institutions had to be registered at a government office, which implied that the government controlled such institutions. In 1982, the Social Education Law was enacted and the control of social education under the law was established. The senior classes in school districts originated as one of the *Samaul Undong* (New Village Movement) community development programmes under the government of President Park Chung-Hee (1971–1979) (Lee, 2008). During this time, the government launched an upgrade of basic rural infrastructure and expanded overall social and economic development in South Korea. The Park administration attempted to reform society through social education and to re-shape the fundamental rules of society (Choi, 2007). Indeed, this movement emphasised a spirit of diligence, self-help and co-operation (Park, 2002), whereby Senior Universities were used to promote and support governmental policies as well as to enhance the spirit of the New Village Movement (Lee, 2008).

Characteristics, Structures and Organisation of the Senior University Movement

At present, there are several providers of educational programmes for older people in South Korea, reflecting the various divisions of responsibilities under the Education and Welfare Law. At a local level, learning programmes are provided through citizen centres belonging to local village offices. Universities and colleges, and higher education institutes, also run lifelong learning centres and provide university-level programmes, while senior and general welfare centres run programmes too. Religious institutions—mainly Catholic, Protestant and Buddhist—are also significant main providers of education for older people. Senior citizens' organisations such as KOPA also play a major role. Of these main providers of older adult education, the Senior Universities of KOPA are most similar to the Universities of Third Age (U3As), although the phrase 'third age' does not appear in their title. Indeed, SUs cater solely for older persons, and their scope is solely for learning. Senior Universities are independent organisations in the voluntary sector and not run by the government, and KOPA is a nationwide organisation.

KOPA differentiates between the terms 'Senior University', 'senior school' and 'senior classes'. While learning classes conducted in the headquarters and regional/provincial association buildings used to be referred to as 'Senior University', those in borough-level centres were called 'senior schools', and the ones in school districts were termed 'senior classes'. Senior schools are nowadays called Senior Universities, and, as noted earlier, the EWL classifies all as 'senior classes'. KOPA organises various activities for senior citizens. It has three founding objectives: to increase senior status in society, uphold senior welfare and promote social relationships among senior people (Choi, 2002). KOPA fulfils these objectives by carrying out the following activities: creating job centres and promoting the enhancement of senior job-related competencies; providing senior leisure facilities; educating on meaningful ageing life; researching and developing policy for senior welfare; the preservation of traditional culture and guidance of adolescents by teaching filial piety; and other activities related to advancing senior people's interests (Kim, 2004). KOPA is headquartered in Seoul, with 16 associations in various cities and provinces. There are also 245 borough-level centres and 64,460 senior community halls (Kyungrodang), situated in villages and urban apartment complexes. In addition, there are 3823 senior classes (senior groups) in every elementary school district. There are 12 senior leader universities in regional/provincial associations and 327 Senior Universities in borough-level centres.

In South Korea, Senior Universities are generally held in buildings of supporting organisations. There are entrance and graduation ceremonies, and the Dean is responsible for the administration of the curriculum. A small monthly or annual fee is paid for membership, while Senior Universities receive financial support from local government as they are registered as one of the senior leisure facilities under the Elderly Welfare Law. Each course runs for more than six months, during which time seniors gather for study and activities once or twice a week. Professional lecturers or course leaders are invited and paid. The contents of the Senior University programmes are predominantly related to aspects of senior life, health, music and dance. Although Senior Universities are run as non-governmental organisations, they can be described as standing on a 'politically-connected model'. The perspectives that have guided the implementation of learning had their origins in politically driven strategy. When KOPA started its operation in 1969, as a popular response to the absence of welfare policies for older people in South Korea, its primary role was to promote the development of such policies. Since South Korea was under non-democratic military rule from 1961 to 1993, the government controlled almost all aspects of public service, including the social-informal learning institutions, thereby focusing the need for campaigners to work more on government-led social and economic development. In Jun's (2014) study, Senior Universities were shown to have developed as part of a politically driven strategy: since 1969, the government significantly enhanced the welfare system to include older people, and KOPA enjoyed the benefits of national welfare policies. The subsequent implementation of learning has been shaped by an organisation-led welfare model, with a formal learning orientation determined by a compulsory set curriculum and top-down management. These features have been reinforced by the Confucian tradition which emphasises vertical, teacher–student relationships in which teachers are regarded as rulers and authority figures.

Senior Universities operate as an adult learning extension of educational provision. In terms of curriculum design, class format and management, centres are patterned on the conventional education system. Indeed, there is a predisposition that each Senior University follows a formal learning system managed by the head quarter of the institution, thus relying upon the pedagogical learning model. Sun and Evans put forward two reasons as to why the pattern of conventional learning is preferred.

> First, South Korea attaches more importance to formal education as a powerful means for social and economic development. Second, being tied to Confucianism there is an attachment to the formal, structured relationship between teacher and student; and having a nationally determined school curriculum is a dominant idea. Confucianism has greatly influenced the educational system of South Korea. Confucianism is deeply concerned with the integration of all human life aspects, especially when it comes to education, civil administration and ceremonies. Confucianism is a part of the South Korean society and is still manifested in the values held even by the new generation of Koreans. Thus, Confucianism and the traditional lifelong educational system in Korea are inseparable….
>
> Jun and Evans (2014: 60)

Indeed, there are stark contrasts between Senior Universities and Universities of the Third Age following the British model. As Jun and Evans argued in a separate article which compares and contrasts these two models,

> First, chairpersons of U3A reported that the economic level of their members ranges from average to very high while SU chairpersons reported the economic level of their members ranges from very low to average. Even though this is not the case for all members, some members of U3A are financially able to invite people to their house and provide cups of tea and biscuits. Their houses are large enough to cater to 5 to 15 members for a group study. In Korea, those who are financially able do not usually attend SU [Senior Universities] or welfare centres because they think SU or welfare centres are for people who have fewer resources. Secondly, on an educational level, U3A members range from average to very high while SU members range from low to average. In most instances within U3A, group leaders are retired academicians, teachers, or professionals so the average level of education is higher when compared to SU members. In Korea, retired academicians or teachers tend to be speakers rather than just members. In SU, there is the concept of teacher and students, not members, and therefore, retired professionals prefer to be a teacher.
>
> Jun and Evans (2007: 69)

The next section, which focuses on the profiles of participants, and the curriculum of Senior Universities serve to illustrate—in practice—some of the above-mentioned differences.

Profile of Participants and Curriculum

Student Profile

According to Jun's (2014) research, conducted in selected Senior Universities in Seoul in 2007, the majority of learners were female (74%). This pattern was also found in Kim's (2017) more recent study of Senior University branches in Daejeon. Table 18.2 presents a comparison of Senior Universities' student profiles in the 2007–2017 period.

Table 18.2   Student profiles in the 2007–2017 period (%)

Variable	Indicator	2007[a]	2017[b]
Gender	Male	25	21
	Female	75	79
Age	61–70	34	
	71–80	62	
	80–90	5	
Home situation	Alone	23	24
	Accompanied	77	
Education	Primary	39	29
Attainment	Middle school	20	29
	High school	28	29
	College/university	10	9
	Other	3	3
Occupation	Housewife	35	
	Education/teacher	19	
	Salaried employee	16	
	Business	14	
	Civil servant	9	
	Manual labour	4	
	Nurse	2	
Financial status	Above average	18	14
	Average	76	75
	Below average	6	10
Health condition	Fairly well	41	46
	Do not feel well but nothing serious	27	40
	Ill	32	14

Source [a]Jun (2014); [b]Kim (2017)

Jun's (2014) study showed that the majority of learners at Senior Universities were aged from 71 to 80 (62%), followed by the group aged from 61 to 70 (34%). In both studies, over 75% of members were still living with family members, and over 40% reported their health condition as 'fairly well'. With respect to educational background, 39% of respondents in Jun's 2007 sample had completed only 'primary level' schooling, compared to 29% a decade later (Kim, 2017). However, in both studies as much as 28% had achieved a high school graduation certificate. A higher proportion had schooling to middle level in Kim's study (29%) compared to the 20% that had middle school certificates. According to both studies, a minority of approximately 10% had college or university-level education. According to Jun's study, 65% of members reported that they had previously been in paid employment, while 35% reported their main occupation as 'housewife'. On financial status, 76% self-reported an average level. Unlike third age universities in some Western countries—notably the UK—Senior Universities do not show tendencies to attract older persons with a high socio-economic status and educational background. Jun and Evans (2014) explained such take-up patterns as reflecting the 'ordinary person culture' of senior classes embedded in community life:

> Even though SUs in South Korea follow the pedagogical model and attempt to shape learning in a formal education style, paradoxically its members are 'ordinary people' who seek non-academic and hobby-centred subjects like dancing and singing in the name of learning and education…Most of the members of SU are female who did not have other jobs except 'being a housewife' in their lives. The outer format of SU has a formal educational institution style but the inner content consists of ordinary people who often do not have enough educational confidence for dealing with academic subjects, or for participating in discussions or sharing their knowledge or skills as is typical for students of [Anglophone and Francophone U3As]….
>
> Jun and Evans (2014: 60)

Indeed, the Senior University model of older adult learning in South Korea is characterised by a self-help model based on the virtues of 'collectivism' and 'obedience', diametrically opposed to the 'individualist' and 'free spirit' character so omnipresent in Western U3As (Jun & Evans, 2014).

Curriculum and Learning–Teaching Strategies

Generally, classes at Senior Universities take place once or twice per week with a fixed time schedule similar to the time schedules found in public schools. Seniors typically sit and listen to lectures by speakers (teachers) and are instructed in structured activities such as music, dance and exercise. Regarding subjects and activities, a metaphor for the Korean Senior Universities could be the 'set menu'. In a cafeteria, customers can choose what they want to eat and pay only for what they eat, and because there is no set menu, customers select a mixture of things they like. However, with a set menu, customers cannot choose their own combination of foods and the set menu price has to be paid even though customers might not want some of the food provided. Senior Universities operate as a course-based initiative in the sense that learners enrol in a particular course which consists of several subjects. Once seniors enter the course, they have to follow the curriculum of that course without selection, generally dominated by lectures on health matters and structured activities of dance, singing and exercise (Jun & Evans, 2014).

In Senior Universities, learning–teaching strategies could be divided into two broad categories: lecture-based and activity-based. In South Korea, discussion is not a preferred teaching–learning method unless it is implemented as part of lecturing. Activity-based learning takes place in practical classes during which students engage in craft-making, dancing, singing, exercising, going on field trips, watching media and observing others. In Jun's (2014) study, 48% of the senior respondents from Senior Universities stated their preferences for 'listening to lectures' over other activities. Classes take place in fixed institutional venues. In EWL, in order to provide a senior class, a lecture room large enough for more than 50 persons is required. In a wide lecture room, students generally listen to lectures and chairs and desks are moved to the back of the hall in case of activity programmes; during these programmes, students dance or exercise together in the room. In case of small group activities, a more formal set up is used in which chairs and desks are set up in a hall. The fixed curriculum was defined as a 'pedagogical model' which focuses on formality of education, reflecting the wider priorities and cultural values in South Korea that attach great importance to formal education as a powerful means for social and economic development (Jun & Evans, 2014). Tutoring plays a limited role and peer learning and teaching are not generally a feature. As noted earlier, in the South Korean SUs, there is a strong distinction between teachers and students rooted in Confucianism, in which teachers are regarded as rulers and authority figures. Normally, speakers or course leaders are invited from outside the university and they are paid. The most popular speakers are university professors.

Active Ageing: Contributions of Senior Universities in South Korea

According to the World Health Organization (2002), active ageing means a continual participation in social, economic, cultural, spiritual and civic affairs. Indeed, older people who retire from work as a result of illness or who live with disabilities can remain active contributors to their families, peers, communities and national life. One of the main purposes for the education of older people is to support the active ageing of older adults in a society. In that respect, SUs in Korea have already contributed to expanding learning opportunities for older people to participate in social, cultural, spiritual or civic affairs. In Korea, the positive impact of Senior Universities on active ageing tends to be evaluated as a concept of life satisfaction and it has been found that participation in Senior Universities is related to high levels of life satisfaction. In the Korean academic database, eight impact studies have been recorded since 2010. Five researches out of eight found that participation in Senior University is related to a high level of life satisfaction and two researchers report a positive effect on self-esteem and ego-integrity, respectively. Hwangbo (2015) examined the relationship between participation in Senior Universities and active ageing and identified contributions in the physical, psychological, social and cognitive domains. In Jun's (2014) study on the benefits that seniors gain from participation in Senior Universities, half of the responses (54/110) recorded 'health' as the key benefit of membership in Senior Universities and then, in order of popularity, friendship, happy mind, confidence, something to do regularly, gaining knowledge, keeping young, pleasure/enjoyment, self-development communication, adaptation to life and the prevention of dementia.

Looking Ahead: The Future for Senior Universities and the Third Age Movement

By 2025, South Korea is expected to become a super-aged society (Korea National Statistical Office, 2016a). The provisions of learning opportunities for old people have become urgent as the number of older people living alone is increasing and so with it is senior suicide. This is because depression and poverty in old age have recently emerged as acute problems in South Korea (Lee & Atteraya, 2018). In 2000, the suicide rate for people aged 65 and over was 35.6 per 100,000, but by 2015, this increased to 58.6. In 2015, the percentage of people living alone aged 65 and over had also increased to 32.9% (Korea National Statistical Office, 2016b). In such circumstances, the expansion of social integration and participation, as well as the involvement of older adults in senior education, is regarded as an effective proactive response. Since the number of older persons with higher-than-average levels of education is rising steadily, more attention is required to meet their educational needs and interests which, in turn, require improvement in the quality of learning programmes targeting this cohort.

According to the *Survey on Living Conditions of the Elderly* (Korea Institute for Health and Social Affairs, 2014), senior citizens who participate in learning programmes provided by citizen centres and public institutions number more than 38% of the older population. Also, while the seniors who participate in learning programmes provided by religious bodies constitute about 9%, the number of participants in citizen centres in local boroughs or districts shows that senior persons find it convenient to access programmes there. With increasing attention to lifelong learning in Korean society, the providers of learning programmes are increasing and diversifying. Therefore, older adults will increasingly find a range of opportunities to meet their learning needs and preferences. Given that the population of older people is more heterogeneous than often assumed, the challenge for Senior Universities is to develop new ways of teaching and learning and programmes that reach out to a wider range of interests and preferences. As increasing number of incoming older adults will be characterised by higher levels of education, it is argued that Senior Universities should take action to develop new programmes and learning approaches, with a focus on active ageing and wider social participation. In Kee's words, despite the great strides in Korea's educational gerontology development, the following areas must be addressed

> …to supply a genuine answer to the rapidly approaching super-aged society: ambiguous division of labor and lack of a unified senior citizen education policy; insufficient financial support, lack of professionalism in teaching personnel; insufficient recognition of the importance of senior citizen education; loose connections between programs and their practical use; failure to address seniors' emotional development; and the separation between the Community for Aged and educational programs.
>
> Kee (2010: 110)

As a precursor to establishing greater variety in learning provision for older people, including Senior Universities, there has to be a change in perspective on the learning of older people. In Korean society, there is an enduring tendency to categorise older people into a group with special needs. In terms of giving special attention to minority groups, this categorisation can be understood to be based on concerns for welfare. It is important to understand that older people have been marginalised and in need of compensation. However, because it is acknowledged that society requires a new triadic conception and paradigm of learning in general—and in this later life context (Aspin, Chapman, Evans, & Bagnall, 2012)—the nature of learning for older people needs to be reviewed as times change. In the evolving discussion of older adult learning, wider perspectives on learning are required to encompass the formal and informal, instrumental and expressive, liberal and occupational, academic and social dimensions (see Fig. 18.1). There is, therefore, a strong case for government to continue to prioritise the least advantaged group of seniors, considering their educational and financial conditions. At the same time, the government could promote self-help and mutual aid among those who are able to organise with less support and less intervention from the government. For the increasingly educated and active groups of older people, the self-help model of British University of Third Age movement is welcomed in South Korea.

Considering that the emergence of third age concepts and debates on educational gerontology and andragogical learning models were active earlier in the UK than in South Korea by several decades, the conditions have now been created for aspects of this collaborative self-help model to be increasingly incorporated into the educational landscape of later life learning in the future.

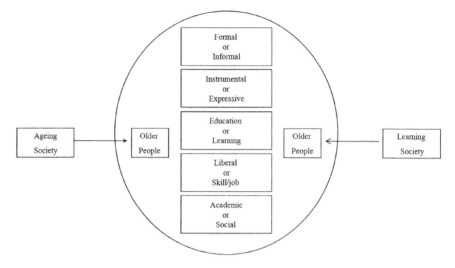

Fig. 18.1 Taking a wider perspective on learning provision for older people in Korea

References

Aspin, D., Chapman, J., Evans, K., & Bagnall, R. (Eds.). (2012). *Second international handbook of lifelong learning*. Dordrecht, Netherlands: Springer.

Byun, J. G. (2004). *Current status of welfare policy in preparation of aged society*. Seoul: Korean Welfare and Health Research Centre.

Choi, D. S. (2002). *The analysis of interest group in the process of senior welfare policy decision making based on Korean Older People Association* (Unpublished Master's thesis). An-dong University, Korea.

Choi, E. S. (2007, November). *Reflections on trends in lifelong educational policies in Korea*. Paper presented at the International Conference on Finding Places for Asian Lifelong Education in Globalizing World, Seoul, Korea.

Han, J. R. (2015). *Educational gerontology*. Seoul: Hakjisa.

Hwangbo, S. (2015). *Effects of Senior University program participation on successful aging of the seniors* (Unpublished doctoral dissertation). Daegu Hanny University, Gyeongbuk, Korea.

Jun, S. K. (2014). *A comparative study of two non-credit educational organisations for older people: The University of the Third Age (U3A) in the UK and the Senior University (SU) in South Korea* (Unpublished doctoral thesis). Institute of Education, University College London, University of London.

Jun, S. K., & Evans, K. (2007). The learning cultures of Third Age participants: Institutional management and participants' experience in U3A in the UK and SU in Korea. *KEDI Journal of Educational Policy, 4*(2), 53–72.

Jun, S. K., & Evans, K. (2014). Learning organisations for older people: Comparing models of learning in the University of Third Age (U3A) in the UK and the Senior University (SU) in Korea. *World Studies in Education, 15*(2), 53–72.

Jung, M. S., Kim, Y. S., & Kim, Y. B. (2014). *Senior education*. Seoul: KNOU.

Kee, Y. (2010). Educational gerontology in Korea: An interpretive and critical study. *International Journal of Lifelong Education, 29*(1), 93–110.

Kim, D. S. (2004). Examining KOPA. *Journal for Bright Ageing, 6*, 8–24.

Kim, S. J. (2017). *Effects of personal variables and participation in Senior University on life satisfaction* (Unpublished master's thesis). Daejeon University, Daejeon, Korea.

Korea Institute for Health and Social Affairs. (2014). *2014 survey on living conditions of the elderly.* Seoul: Korea Institute for Health and Social Affairs.

Korean Older People Association. (1989). *The 20 year history of KOPA.* Seoul: Korean Older People Association.

Korea National Statistical Office. (2005). *Future population special projection.* Daejeon: Korea National Statistical Office.

Korea National Statistical Office. (2016a). *Future population projection: 2015–2065.* Daejeon: Korea National Statistical Office.

Korea National Statistical Office. (2016b). *Statistics on older persons.* Daejeon: Korea National Statistical Office.

Lee, H. Y. (2008). *A study on training policy for professionals of elder education in Korea* (Unpublished master's thesis). Seoul National University, Korea.

Lee, S. Y., & Atteraya, M. S. (2018). Depression, poverty, and abuse experience in suicide ideation among older Koreans. *International Journal of Aging and Development.* http://journals.sagepub.com/doi/abs/10.1177/0091415018768256. Accessed 29 June 2018.

Park, J. G. (1999). *Wisdom of elderly life.* Seoul: Korean Institute of Gerontology.

Park, S. J. (2002). The change of South Korea adult education in globalization. *International Journal of lifelong Education, 21*(3), 285–294.

World Health Organization. (2002). *Active ageing: A policy framework.* Geneva: World Health Organization.

Soo-Koung Jun holds the post of Assistant Professor at Namseoul University in South Korea. She achieved a B.A. in education from Chungnam National University and an M.A. in education as well as from Seoul National University. She was awarded a Ph.D. in 2014 under the supervision of Prof. Karen Evans at the Institute of Education, University of London. Her Ph.D. thesis was titled *A comparative study of two non-credit educational organisations for older people: The University of the Third Age in the United Kingdom and the Senior University in South Korea.* Her research interests are educational gerontology and professional learning.

Karen Evans is Emeritus Professor of Education at UCL Institute of Education. Formerly Chair in Education and Lifelong Learning at UCL and Head of the School of Lifelong Education and International Development, she specialises in international and comparative studies of education and learning throughout the life course. She is currently Chair of the Asia-Europe Lifelong Learning Hub Research Network on work and learning, and Editor of the Springer Lifelong Learning Series, and previously, she jointly edited the journal COMPARE. She is also Fellow of the Academy of Social Sciences and recipient of the 2017 European Commission's Vocational Education and Training Researcher Award.

Chapter 19
From Social Welfare to Educational Gerontology: The Universities of the Third Age in Taiwan

Shu-Hsin Kuo and Chin-Shan Huang

Introduction

The development of science and technology is continuously increasing human life expectancy. In 1993, with more than 7% of its population being over the age of 65, Taiwan fit the United Nations' definition of an ageing society. In March 2018, more than 14% of Taiwan's population was aged 65 years or older. According to the National Development Council (2016), Taiwan will become a 'super-aged' society by 2026, with more than 20% of its population expected to be over the age of 65. An aged society faces multifaceted challenges which require more social and health care services and assistance. One key challenge is to provide older adults with educational and learning opportunities—which can be offered through Universities of the Third Age (U3As)—to enable them to age actively and successfully. As this chapter attests, the development of U3As in Taiwan can be divided into two temporal stages, from 1978 to 2008 and post-2008.

1978–2008: The 'Social Welfare' Stage

The Evergreen Club—the first learning organisation for older adults in Taiwan—was established in 1978; its objective was to systematically provide older persons with educational activities. The aim of this was to call for society to improve its levels of respect towards older persons, and to be more sensitive to their well-being.

S.-H. Kuo
Toko University, Puzi City, Taiwan
e-mail: gosuhsin@gmail.com

C.-S. Huang (✉)
National Chung Cheng University, Chiayi, Taiwan
e-mail: aducsh@ccu.edu.tw

© Springer Nature Switzerland AG 2019
M. Formosa (ed.), *The University of the Third Age and Active Ageing*, International Perspectives on Aging 23, https://doi.org/10.1007/978-3-030-21515-6_19

Meanwhile, The Evergreen Club also sought to promote leisure activities for older persons to improve their quality of life, with particular emphasis on their physical and mental health. Activities included lectures, seminars, vocational programmes and entertainment activities that combined leisure with socialisation and learning. In December 1982, the Social Affairs Bureau of the Kaohsiung City Government collaborated with the Young Women's Christian Association to jointly launch a university for older adults: Chang Ching Shyue Yuan (CCSY). Between 1983 and 1993, older adult education in Taiwan was mainly led by CCSYs established by social service departments of local governments. In October 1987, the Ministry of the Interior (MOI) launched a plan to establish CCSY in the whole Taiwan Province—thus requesting social service departments in all counties and cities to select appropriate venues for the setting up of older adult universities. As of 2011, all of the 21 counties and cities in the Taiwan region, with the exception of Lienchiang County, included a CCSY, bringing the total number of CCSYs to 341; this meant that there were 4226 classes totalling 134,058 participants (Ministry of Interior, 2012). However, in 2008 the issue of late-life learning had not yet attracted the attention of the public education department in Taiwan, and activities and classes—mostly focusing on leisure and entertainment—were led by the MOI.

Post-2008: The 'Educational Gerontology' (Active Ageing Learning) Stage

In 2006, the Ministry of Education (MOE) in Taiwan enacted legislative reforms to release a white paper, entitled *Towards an aged society: Policies on education for older adults*; the aim of this was to ensure that older persons' rights to engage in learning opportunities are respected, thus ensuring their successful and healthy ageing. This was the first clear declaration—as well as the most comprehensive plan yet—in favour of adult learning in later life in Taiwan. Given that the MOE had begun to recognise the importance of education for older adults and to proactively establish learning units for older persons, an Active Ageing Learning Center (AALC) was established in 2008 by the Department of Adult and Continuing Education of the National Chung Cheng University (NCCU). It was eventually commissioned by the MOE to assist local city and county governments to set up AALCs and to organise active ageing learning activities, thus effectively launching the development of educational gerontology in Taiwan (Ministry of Education, 2016a).

In 2008, the MOE launched AALCs in every town to encourage higher numbers of older adults to participate in learning activities (Huang, 2010). Presently, AALC programmes are primarily educational and are usually combined with existing learning programmes organised by senior centres, non-profit organisations, community development associations and schools. The AALCs target citizens aged 55 and over, for in-place learning, by means of an AALC in various towns and villages. A total of 104 AALCs were established in 2008, with this number reaching 339 by 2016

(Ministry of Education, 2016b). In recent years, a total of 1766 community learning centres were established in 988 villages (Huang, 2016). In addition to AALCs, the first U3A in Taiwan—or as it is referred to in the country's native language, *Le Ling Da Shyue*—was established in various universities across Taiwan:

> …the Ministry of Education launched a university-based programme in 2008. This programme offers a short-term or Elderhostel approach to learning activities on college campuses, called *Le Ling Da Shyue*. The idea for this learning approach was derived from the Elderhostels in the United States. Older adults who participate in this programme can access academic resources and have opportunities to interact with university students. Also, the classes in this programme can utilize the facilities and teachers from the universities. In 2012, there were 87 colleges and universities taking part in this programme, compared to only 13 universities in 2008.
>
> Lin and Huang (2016): 427

However, because of the unique historical, socio-economic and educational context of Taiwan, Taiwanese UOAs are rather different to U3As in France, Britain or other countries:

> First, Taiwanese UOAs, unlike British U3As, tend to be established and managed by local governments. Second, Taiwanese UOAs, unlike French U3As, tend to run their own activities without any assistance from local universities or colleges. Finally, the instructor-centered lecture is the most often used by the teachers in the classrooms of Taiwanese UOAs. This feature of Taiwanese UOAs is very similar to that of French U3As. Nevertheless, Taiwanese UOAs remain different than French U3As because they are not connected to local universities. Hence, Taiwanese UOAs form a unique model, based on the different historical, socioeconomic, and educational context of Taiwan.
>
> Huang (2005): 513–4

In 2008, a programme was launched to offer a short-term Elderhostel approach to learning activities on college campuses where older adults could access academic resources and interact with younger university students. A total of 13 universities held these five-day Elderhostel-approach programmes, and 820 people aged 60 and over participated in them. In 2010, Le Ling Da Shyue was launched to offer learning programmes to citizens over the age of 55. Programmes were divided into two 18-week-long semesters every year, of nine study units per week. Through the efforts of the MOE—and all public and private universities and colleges—the number of Le Ling Da Shyue doubled over a six-year interim period: from 56 in 2010 to 108 in 2016. The growing number means that older adults are being offered more opportunities and channels to share the educational resources, and that their learning rights are better protected (Ministry of Education, 2016b). As shown in Table 19.1, until 2008 the development of U3As in Taiwan can be divided into two stages.

Indeed, the approach to Taiwan's U3As has shifted from an MOI-led social welfare one to an MOE-led educational gerontology approach. In terms of institutes and courses, they have also moved from the CCSYs of the social welfare system to the AALCs within the educational gerontology system. The former focus on leisurely and entertainment-oriented activities, which have existed in Taiwan for more than 30 years. However, as less pragmatic arguments have been presented in favour of this approach, this article will not elaborate on it. The latter—the educational gerontology

Table 19.1 Course design and plan for cultivating professionals for AAL in Taiwan

Period	Social Welfare 1978–2008	Educational gerontology 2008
Governmental department	Ministry of interior	Ministry of education
Important policies	Implementation directives for establishing Chang Ching Shyue Yuan in the Taiwan Province (1987)	The white paper of *Toward an aged society: policies on education for older adults* (2006)
Learning Institutes	Chang Ching Shyue Yuan Universities for aged adults	Active Ageing learning centres Le Ling Da Shyue
Courses	Leisure activities for aged adults to improve the quality of their lives as well as their physical and mental health	Developing AAL core courses, courses initiated by individual centres and courses of social service with the goal of active ageing

approach—is based on theories and the pragmatic guidance of relevant scholars in the Department of Adult and Continuing Education and the Ageing and Education Research Center of the NCCU, and is therefore derived with an educational approach in mind; as a result, it has resulted in the development of many course structures and activities for cultivating lecturers. Implementing AALCs is about establishing goals and visions, the professional training of staff, and the design and arrangement of learning activities to gradually reduce the proportion of hobbies and leisure activities, year by year. In addition, it also emphasises the training and professional growth of AAL lecturers with the assistance of the head guidance group and the district guidance groups, aiming to enhance AAL lecturers' understanding of aged adults' specific physical and mental needs, and to strengthen their capabilities towards course planning and design, and mostly to improve their overall teaching efficiency.

Course Design in AALCs

Since 2008, AAL education has been well developed in Taiwan and the MOE-led AALCs are different from the CCSYs led by social welfare departments. In terms of course design, AALC courses differ from previous CCSYs in that they are more organised and systematic. Firstly, AALC course design is based on active ageing theory and McClusky's (1974) Margin Theory of Needs, offering AAL core courses by scholars (Wei, Chen, & Lee, 2014). Secondly, it encourages AAL service clubs to provide voluntary services through social services so as to bring learners' value into full play. Thirdly, in addition to courses and activities held in AALCs, the existing centres are required to incorporate into their course plan at least eight-hour activities for each of the villages. With the assistance of existing lecturers and volunteer workers, AALCs should launch AAL subdivisions at the communities in their neighbourhood in an organised and systematic manner. These three types of courses are described hereunder.

AAL Core Courses

AALCs differ from traditional education for aged adults in that each AALC is required to include AAL core courses in its course design (Wei, 2012). The course, focusing on active ageing, can be divided into five categories with 39 subjects. As shown in Fig. 19.1, every AALC can develop courses with characteristics of its own to assist aged adults in better understanding the purposes and ideals of promoting the educational gerontology policies and better preparing themselves for the ageing lives (Ministry of Education, 2017a).

The hours required for the AAL core courses by the MOE are 624 h for AALC Demonstration Centres, accounting for 40–50% of the total hours; 520 h for *Quality A Centres*, accounting for 35–50% of total hours; 312 h for existing centres; and 208 h for newly established centres, accounting for 30–40% of the total hours (Ministry of Education, 2016d).

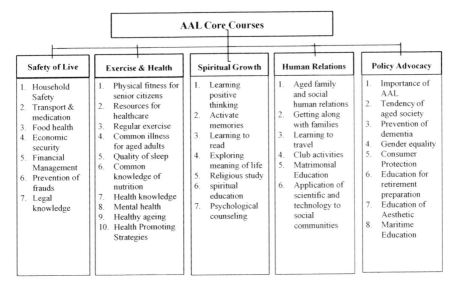

Fig. 19.1   AAL Core courses

AAL Courses of Service

AALCs are required to provide social service courses and encouraged to establish AAL service clubs with the goal of using what is learned in AALCs to serve society. For example, members of the Flute Club at the AALC in Shuishang Township of Chiayi County took the initiative to learn to play the instrument as well as to put what they have learned into use to serve society. Arranged by the AALC, they regularly participated in performances at weekend music markets, charity institutes and hospitals and played more than 10 performances around various villages in the township, thus creating a recycling value of 'aged people serving aged people' (Hou, 2013).

AAL Extension Courses in Communities

In order to better explore and deliver the resources of education for aged adults, particularly for those who reside in remote rural areas or who are physically challenged, the MOE requires that all existing AALCs make extension plans, asking the teams in existing AALCs to incorporate into their course plan at least eight-hour activities for each of the villages in addition to the courses and activities held in the AALCs. With the assistance of existing lecturers and volunteer workers, AALCs are able to launch AAL subdivisions in the communities, in an organised and systematic manner (Ministry of Education, 2016b).

Plan for Cultivating AALC Professionals

AALCs are organised and managed by different governmental departments and civil organisations. AAL professionals, such as administrators, course planners and lecturers, are responsible for AAL quality and efficiency. In April 2012, the MOE promulgated Directives for Cultivating Professionals for AAL Education, aiming to enhance the professionalism of AAL education staff, as well as to redevelop and utilise the human resources of older adults. The training courses ensured an upgrade of skills of AAL professionals, namely AAL lecturers, supervisors, organisers in competent authorities and leaders of active learning groups. The plans for cultivating AAL professionals—from 2012 to 2017—are further outlined below.

Training of Professionals for AAL Education at Different Stages

The aim of the Directives is to upgrade the skills of AAL professionals. The training project is divided into three stages for a total of 126 h, comprising of 36 h of basic training, 42 h of advanced training, 34 h of practice and 14 h of seminars and presentations. A certificate of competence is conferred on those who have completed all training courses. The contents of the basic and advanced training courses differ in that the basic training focuses on common courses such as aged society and knowledge about ageing, whereas the advanced training provides different courses for AAL lecturers and leaders of active learning groups. Training courses include:

- *Basic Training Courses.* Common study units in five major categories, including aged adults learning and active ageing, such as the current situation of aged society, policy for aged adults learning, active ageing learning and gender issues; important concepts about physical ageing; important concepts about mental ageing; important concepts about social ageing; and project planning and service ethics.
- *Advanced Training Courses for AAL Lecturers.* Study units which include topics such as characteristics and needs of aged adults learning, basic concepts about teaching aged adults, design and compilation of materials for teaching aged adults, methods and techniques of teaching aged adults, interaction and practice of teaching aged adults, leading activities for aged groups, and resources of teaching aged adults and multimedia application, etc.
- *Advanced Training Courses for Leaders of Active Learning Groups.* Learning programmes that cover topics such as basic concepts about active learning groups, development and operation of active learning groups in Taiwan and abroad, planning and management of active learning groups, leading techniques and resources exploration for active learning groups, marketing and development of active learning groups, etc.

Plan for Cultivating AAL Planners

In 2014, the NCCU Ageing and Education Research Centre, commissioned by the MOE, launched the AAL planner project, aiming to align the establishment and management of AALCs with regional features and to develop learning models that comply with the needs of aged adults in local communities. The project targeted AALC staff nationwide, particularly incumbent AALC directors, teachers and volunteer workers, with extra emphasis on those with at least two years of experience. Participants in the project were recommended by AALC directors and then selected by the organiser. The aim of the project was to promote a new teaching model, known as 'Teach 123', in which '1' means one main point to be explored in each teaching unit; '2' represents two teaching activities related to the main point in each unit, including such experiences as waking up consciousness and knowledge-based teaching strategies; and '3' denotes three action strategies that learners have to apply to real life after the end of each unit (Wei, 2016).

Plan for Cultivating a Senior Planner

In 2016, the 'Senior Planner' project was launched; this was an advanced training project for AAL planners focusing on 'activating memories and recognising dementia' to cultivate seed lecturers for AAL core courses. It is expected that those 'senior planners', who have completed the training courses, are able to assist newly established AALCs and AALCs in remote rural areas with disseminating knowledge on dementia. The project's goal is to cultivate advanced planners capable of developing and implementing diversified and innovative activities for aged society. Participants are to be capable of designing a 'Teach 123' plan for recognising dementia. Such plans should include topics as basic concepts on AAL core courses, plans for recognising and preventing dementia, and Standard Operating Procedures (SOP) for courses of activating memories, teaching practice for such courses and so on. Eventually, a total of 54 participants completed the entire course and submitted their teaching achievements. They subsequently held 237 lectures to a total of 4771 audiences (23% male and 77% female) (Wei, 2016).

Plan for Cultivating Lecturers for AAL Core Courses

The design of the 2017 training course for AAL professionals was based on the concept of creating effective teaching and learning. The goal was set for AALCs and institutes related to education for aged adults to conduct AAL core courses with the assistance of AAL core course planners who have completed the relevant training. The training course was divided into four stages: recruiting and selection, training and learning, developing teaching plans and teaching practice. Among them, the training and learning accounted for 7 h, after which participants were expected to develop a teaching plan for 16 h AAL core courses based on their content. After that, they were required to conduct 16 h of teaching practice in AALCs or other institutes and finally to complete a self-evaluation form and collect such information as the effects of learners. Certificates of accomplishing training hours would only be conferred on those who had participated in all training activities at all stages and had submitted completed materials (Ministry of Education, 2017a). Since embarking on the educational gerontology stage in 2008, MOE-led AAL education has taken to course planning in a more organised and systematic manner, which has helped learners achieve their goals through AAL core courses and encouraged more people to participate in AAL by means of combining practical services to extensions of the communities. Furthermore, by providing training courses for AAL lecturers, course planners and leaders of active learning groups at different stages, it has helped to enhance their knowledge of aged adults learning. This does not only strengthen their professional capacities, but also help promote service quality to, and learning efficiency of, learners in the following ways:

- *Service quality of AALCs.* Since their establishment in 2008, AALCs have been established in various villages and townships in Taiwan. The article will further evaluate their service efficiency from such aspects as the extension to villages, management and implementation, a survey of AALC participants' needs for learning and re-employment, and their gender variation in order to better understand the AALC service efficiency.
- *Efficiency of AALC management and implementation.* AALC participants are generally satisfied with courses and teaching staff. Satisfaction does not generally depend on whether AALCs charge fees or not, and AALCs therefore have the potential to launch courses that charge users (Ministry of Education, 2016c). It is recommended that the head guidance group and district guidance groups continue assisting AALCs to develop new courses, such as training courses and regular meetings, to upgrade the quality of courses and to meet participants' needs.
- *Efficiency of AALC extension courses in communities.* While most of the AALC managers support extension courses to deliver education resources to disadvantaged areas, nearly 70% of AALCs encounter difficulties during the process, most common of which are lack of manpower and lack of venues in that order (Ministry of Education, 2016c). Therefore, AALCs should make good use of the manpower of volunteers and establish a database of talents for extension courses in communities so as to increase its human resources. In addition, they should actively seek out venues for holding courses, such as neighbourhood activity or religious centres.
- *AALC participants' needs for learning and re-employment.* Currently, AALC participants need knowledge in living capability more than they do in re-employment—only around 30% of them show any interest in re-employment (Ministry of Education, 2016c). Therefore, it remains to be evaluated whether AALCs are to include the topic of re-employment in their courses in the future. It is also possible to consider the need for learning skills targeted at non-fixed term employment of flexible working hours.
- *Gender difference of AALC participants.* Statistics in 2010 showed three times as many female participants as there were male participants (Hu, 2011). Hsiao, Wei, Lee, and Chen (2013) conducted a large-scale survey on the participants in 209 AALCs, 56 U3As (Le Ling Da Shyue) and 322 CCSYs. With a total of 3179 returned questionnaires, the study shows that in terms of gender, female participants exceed male ones in numbers; in terms of their perception of the importance of AAL subjects, female participants considered subjects such as exercise and health, spiritual growth, human exchanges and service, to be most important; females also outnumbered males in AAL course participation.

In spite of the fact that AALCs are short of manpower and venues for extension courses in communities, most AALC managers support further extension to accessibility of the AAL; AALC participants need more knowledge in living capability than in employment, indicating that most of them are probably retired and do not take part in the courses for the purpose of employment; in terms of participants' gender, males are far less likely to enrol than females, and differences were also found to exist in their perception of learning subjects. Therefore, AALCs should provide different subjects and contents in line with the appropriate development of different genders in order to meet their different needs.

Characteristics and Future Prospects of U3As in Taiwan

Education for aged adults in Taiwan, whether it is long-established CCSYs or lately launched AALCs, follows a basic operation model, and both share a lot of similarities. For instance, both of them rely on government funding, hire professional lecturers and maintain the same classroom-learning model. However, in comparison with the courses focusing on leisure and entertainment provided by the social welfare-oriented CCSYs, AALCs construct the AAL core courses with a theoretical approach to promoting active ageing; through guidance groups composed of teachers from departments related to adult education in universities and colleges, AALCs ensure the quality and service efficiency of AAL by providing training courses for professionals, evaluation indicators and on-site visitations. The characteristics and future prospects of U3A in Taiwan are further elaborated on as follows.

Financially Supported by Central Government

Taiwan's AAL education began with the promulgation of the Policy White Paper, *Toward an Aged Society: Policies on Education for Older Adults* by the central government department (MOE) in 2006. Since then, using government budgets, the MOE has commissioned the Department of Adult and Continuing Education of the NCCU to take on the role of AAL head guidance group; it then commissioned departments related to adult education in four other universities to act as district guidance groups. It also subsidises local governments to establish AALCs in various villages and townships and to hold training courses for AAL professionals at different levels and categories through guidance groups so as to enhance professionals' knowledge of active ageing, as well as their efficiency and effectiveness in teaching strategies (such as AAL planners' Teach 123 strategy) to promote learning and service efficiency for learners.

Constructing AAL Core Courses with a Theoretical Approach

Scholars plan AAL core courses based on active ageing theory and McClusky's (1974) Margin Theory of Needs (Wei, Chen, & Lee, 2014). The head guidance group then designs the AAL core courses for AALCs accordingly (Ministry of Education, 2017b) and encourages individual AALCs to develop courses with unique characteristics of their own. The required hours for AAL core courses account for 30–50% of the overall hours. Moreover, AALCs are also required to provide courses in social service and encouraged to establish AAL service clubs so as to satisfy the need for self-actualisation, the highest in Maslow's hierarchy of needs.

AALCs Participated by Both Genders

By analysing the gender difference of AALC participants, it is found that female participants participate more than male ones, and participants of different genders also have different perceptions of the importance of the learning subjects. Female participants tend to prioritise subjects such as exercise and health, spiritual growth, human exchanges and service. Therefore, U3As should offer more flexibility in its course design, course content and teaching methods to provide different learning subjects and contents in line with the appropriate development of different genders in order to meet their different learning needs.

Returning to the Self-learning-Oriented Approach with Learners as the Centre

As the number of older adults born after World War II increases, their level of education is also increasing. Thus, further consideration needs to be given to whether the traditional teaching model for children is still applicable. Currently, AAL training provided by U3As is exclusively for AAL lecturers, supervisors and organisers in competent authorities and leaders of active learning groups; there is neither a campaign nor training provided for AALC participants. As a result, participants are passive learners without an active initiative. The well-known University of the Third Age in the UK does not receive government subsidy, and the main funding is shared by participants (Formosa, 2014, 2016). It does not employ professional lecturers to teach the classes, but allows participants to share their experiences with each other. It no longer operates a classroom teaching model, where lecturers teach and students learn, but a model of peer coaching. The participant-oriented learning model of the West should be introduced to Taiwan's AAL education in the future to allow more space for participants to take part and engage in discussion, so as to genuinely enter into the model of adult education.

Conclusion

The development of U3As in Taiwan was affected by the direction change of policies in central government and can be distinguished into two stages: pre- and post-2008. Before 2008, it was CCSYs established by social welfare system that provided older persons with activities and courses focusing on leisure and entertainment. After 2008, AAL from the educational gerontology system gradually took shape in Taiwan, developed the AAL core courses and professional training for aged adults and has thus provided Taiwanese citizens aged 55 and over with AAL channels to reach the goal of active ageing.

For the last decade, AALCs have accepted financial support from central government and review indicators designed by guidance groups and have participated in relevant training courses for professionals, evaluation of AALCs' efficiency of implementation and on-site visitations, thus contributing to shaping the unique characteristics of AAL in Taiwan, which has taken roots in the first line of education. It is, however, noteworthy that AAL education has always been practice oriented, focusing on practical activities without promoting AAL academic research. In future, U3A in Taiwan should encourage and support the convening of relevant academic conferences, such as international conferences or seminars on U3A, to facilitate exchanges between—and visits to—AAL education institutes in Taiwan and abroad. By sharing each other's experiences, it is possible for us to learn from each other and to stimulate innovative thinking for the development of Taiwan's U3A in the next decade, as well as the development of AAL education as a whole. Finally, Taiwan's AALCs should aim for a 'joint learning of all people' to promote the AAL era where people of all genders are willing to take part. Meanwhile, they should also encourage older adults to deliver and give full play of their 'silver' strength through learning across generations and service courses so that Taiwan—a country marching into the super-aged society—may offer a friendlier and more appropriate living and learning environment without age discrimination.

References

Formosa, M. (2014). Four decades of Universities of the third age: Past, present, and future. *Ageing & Society, 34*(1), 42–66.

Formosa, M. (2016). Malta. In B. Findsen & M. Formosa (Eds.), *International perspectives on older adult education: Research, policies, practices* (pp. 161–272). Cham, Switzerland: Springer.

Hou, C. C. (2013). *Active Ageing-ing*. Taipei: Ministry of Education.

Hsiao, Y. F, Wei, H. C, Lee, Y. H., & Chen, G. L. (2013). *Study on the active ageing learning needs of participants of different genders in the education for aged adults*. Paper presented at the First Annual Conference of Taiwan Association of Gerontology and International Conference, Republic of China.

Hu, M. C. (2011). *Active ageing learning in Singapore: Organization and practice*. Kaohsiung: Liwen Publishing Group. (Chinese).

Huang, C. S. (2005). The development of a University for older adults in Taiwan: An interpretive perspective. *Educational Gerontology, 31*(7), 503–519.

Huang, C. S. (2010). The development of educational gerontology in Taiwan: An interpretive and critical perspective. *Educational Gerontology, 36*(10–11), 968–987.

Huang, Y. L. (2016). Policy of education for aged adults under the ageing population. *Public Governance Quarterly, 4*(1), 106–114.

Lin, Y., & Huang, C. (2016). Taiwan. In B. Findsen & M. Formosa (Eds.), *International perspectives on older adult education: Research, policies, practices* (pp. 421–431). New York: Springer.

McClusky, H. Y. (1974). Education for ageing: The scope of the field and perspectives for the future. In S. Grabowski & W. D. Mason (Eds.), *Learning for ageing* (pp. 324–355). Washington, D.C.: Adult Education Association of the USA.

Ministry of Education. (2006). *Towards an aged society: Policies on education for older adults.* Taipei: Ministry of Education.

Ministry of Education. (2016a). *What is "Active ageing learning"?* http://moe.senioredu.moe.gov.tw/UploadFiles/20160423113736250.docx. Accessed August 20, 2018.

Ministry of Education. (2016b). *Ministry of Education explanation of policy for active ageing learning.* http://moe.senioredu.moe.gov.tw/UploadFiles/20160727094441933.pdf. Accessed August 20, 2018.

Ministry of Education. (2016c). *Report on the survey plan for service efficiency and needs of active ageing learning nationwide.* http://moe.senioredu.moe.gov.tw/UploadFiles/20160905055014625.pdf. Accessed August 20, 2018.

Ministry of Education. (2016d). *Principles for writing plan of active ageing learning centers for applying for the MOE subsidy to municipals and counties (Cities).* http://moe.senioredu.moe.gov.tw/UploadFiles/20161024072638610.pdf. Accessed August 20, 2018.

Ministry of Education. (2017a). *Pilot Plan for promoting life-long learning of elderly active learning groups in 2017.* http://moe.senioredu.moe.gov.tw/UploadFiles/20170216111226516.doc. Accessed August 20, 2018.

Ministry of Education. (2017b). *MOE plan for the training of planners for the active ageing learning cores course.* http://moe.senioredu.moe.gov.tw/UploadFiles/20170210095611545.pdf. Accessed August 20, 2018.

Ministry of Interior. (2012). *Annual Statistics of Welfare Service for the Elderly in 2011.* http://sowf.moi.gov.tw/stat/year/y04-16.xls. Accessed August 20, 2018.

National Development Council (2016). *Estimation of Population in the Republic of China: 2016–2061.* http://www.ndc.gov.tw/Content_List.aspx?n=84223C65B6F94D72. Accessed August 20, 2018.

Wei, H. C. (2012). *Active ageing learning in Taiwan.* Taichung: Wu-nan Books Inc. (Chinese).

Wei, H. C. (2016). Plan for cultivating active ageing learning planner: Design of teach 123 model. *T&D, 221,* 1–22.

Wei, H. C., Chen, G. L., & Lee, Y. H. (2014). Structure and assessment of active ageing learning courses in education for aged adults. *Chung Cheng Educational Studies, 13*(1), 45–87.

Shu-Hsin Kuo holds the post of Assistant Professor at the Department of Social Work at Toko University, Taiwan. She is also an academic staff of the Department of Adult and Continuing Education at National Chung Cheng University. She obtained her Ph.D. in 2016 from National Chung Cheng University, in which she was awarded the Doctoral Thesis Award from the Ministry of Science and Technology. In 2017, she was Social Work Supervisor at the local government unit of Chiayi County. In 2018, she was awarded the Outstanding Performance of Social Work Supervision Award. Her research interests include educational gerontology, active ageing and social welfare services for older persons.

Chin-Shan Huang holds the post of Professor at the Department of Adult and Continuing Education at National Chung Cheng University, Taiwan. From 2015 to 2018, he was the Director of the Department of Adult and Continuing Education. He is the editor for the Journal of Adult and Lifelong Education in Taiwan and also a council member of Taiwan Association of Gerontology. His main research interests are focused on learning in later life, ageism studies, ageing education and intergenerational learning.

Chapter 20
'Lifelong Education' Versus 'Learning in Later Life': A University of the Third Age Formula for the Thailand Context?

Cameron Richards, Jittra Makaphol and Thomas Kuan

Introduction

In a report by UNESCO's (2016) lifelong learning division, Thailand was named as one of a small group of countries where 'third age education' was especially important in national adult education policy. This is in part due to how Thailand has become yet another 'ageing society' (Chaitrong, 2017). Indeed, projected current and future rates of decline in the working age population are reported to be higher in Thailand than in any other developing East Asian and Pacific country (World Bank, 2016). Thailand's strong non-formal education policies and practices that include the older population who grew up in rural areas and never previously had a chance at formal education are another factor contributing to this situation (DW News, 2018). Moreover, Thailand is a predominantly Buddhist society, a developing country where traditional cultural values and knowledge remain important to the wider community.

At face value therefore, one might expect Thailand to be particularly receptive to the U3A concept. But a previous study has shown that there was some general resistance to this movement, especially by people without formal education who tended to associate the term with higher education institutions, and who resultantly felt intimidated by this association (Ratana-Ubol, Charungkaittikul, & Sajjasophon, 2012). Few Thais are aware of the existence of the British model of the U3A that had moved away from the earlier French connections with established universities (and

C. Richards (✉)
Southern Cross University, Lismore, Australia
e-mail: cameronkrichards@gmail.com

J. Makaphol
Silpakorn University, Bangkok, Thailand
e-mail: makapol_j@su.ac.th

T. Kuan
U3A Singapore, Singapore, Singapore
e-mail: kuanthomas@gmail.com

© Springer Nature Switzerland AG 2019
M. Formosa (ed.), *The University of the Third Age and Active Ageing*, International Perspectives on Aging 23, https://doi.org/10.1007/978-3-030-21515-6_20

academic tendencies) to exemplify informal and collaborative notions of lifelong learning for seniors more consistent with local needs and practices (Formosa, 2014). As a result of their influence in Thai society, local university support for some kind of variation of the British (not French) model might be the key to promoting both (a) the U3A as a specific concept (as distinct from say related models such as China's 'universities for the elderly') and (b) related semi-organised or community-based encouragement of lifelong learning centres for seniors reasonably consistent with established notions of 'third age education' (Ratana-Ubol & Richards, 2016).

The chapter explores some of the interesting global and local implications of a new initiative, by a university lifelong education department, aiming to promote the U3A concept in Thailand in terms of a partnership collaboration with both the local community and by networking with groups of older persons in other areas and provinces. This initiative will be split into three sections. The first examines the influential Thai context of relatively strong non-formal education policies and practices. The second discusses the plans and aspirations of Silpakorn University's Department of Lifelong Education to become a centre for promoting the U3A movement in Thailand in terms of an adapted model of collaboration with a local community. The last section explores the 'Silpakorn model' as an interesting focus on some dilemmas in relation to lifelong education, as both a policy and practical focus of lifelong education and learning around the world.

Lifelong Education Practices in Thailand and Their Relevance for 'Third Age' Citizens

Like many other countries around the world, Thailand is ageing rapidly (Jitapunkul & Wivatvanit, 2009). A recent National Economic and Social Development Agency report projects that Thailand will become a 'fully fledged' ageing society by 2021 (Chaitrong, 2017), which is when the percentage of citizens aged 60 and over is projected to reach 20% of the overall population. It is promising that in Thailand various segments of local society—including government, non-government organisations, universities and families—are increasingly interested in providing better support to seniors to achieve sufficient levels of competency and security, enabling them not only to survive but to contribute to a future society. In contrast to typical OECD ageing societies, non-formal and informal modes of lifelong education for older persons in this country are linked to traditional notions of 'Thai local wisdom' (Jungck & Kajornsin, 2003).

The commitment to quality education in the national Constitution of Thailand recognises that formal, non-formal and informal learning are all aspects of related policies to support lifelong learning for all (Thailand Constitutional Court, 2007). The 2008 *Promotion of Non-Formal and Informal Education Act, B.E. 2551*, Sect. Towards an 'optimal' U3A model of seniors' education and learning?, directs that 'all Thai people shall receive non-formal and informal education to develop knowledge, skills and aptitude throughout their lives'. On closer inspection, this is held to involve four related lifelong learning strategies: (i) general non-formal education provision including literacy learning equivalent to formal school programs; (ii) vocational education programs (including both short and long courses in skills training and professional development); (iii) life skills and personal development courses; and (iv) a 'learning communities' approach to social development and inclusion activities.

Currently, these strategies of non-formal and informal education, and lifelong learning, are focused on the Thai community college system which was adapted from a similar US model (Intarakumnerd, 2012). The Thai community college system is typically linked to local community needs (e.g. tourist guide training in colleges in Tourist areas like Phuket). Whilst community colleges include associate degrees and certificate programmes, older persons are especially catered for in terms of both 'remedial education' and also continuing education programs conducted in these colleges (Charrurrangsri, 2012). Older persons living in rural areas, especially those who did not get the chance of a formal education when younger, are a key focus of literacy programmes—which also include others such as hill tribal people, migrants and street kids. The continuing education programmes catering for older persons overlap with a potential key preserve of the U3A in that they typically involve short courses in areas such as traditional culture and art appreciation, and on functional needs such as health, financial literacy and access to computers and the Internet.

Health is a particularly interesting case in point. A key finding of an earlier study was that there was general recognition by older citizens and policy makers that the major lifelong learning goal for Thai senior citizens was 'health'—in that lifelong education policies and practices should 'help senior citizens stay so healthy that they will enjoy relationships with their families and are able to happily adapt themselves to the communities and society' (Ratana-Ubol et al., 2012: 3). The focus on health may interestingly involve either traditional local wisdom or modern health knowledge. Thus, both traditional knowledge and public health practitioners are regularly involved in providing short-term courses at community colleges and elsewhere including ones on sanitation, mental health and adaptation to society.

The Thai approach to the fourth strategy (that is, a 'learning communities' approach to social development and inclusion activities) is particularly noteworthy. The most significant and widespread instances of this are the 'sufficiency economy' development projects inspired by the 'sufficiency economy philosophy' of the late King Bhumibol. This particular philosophy promotes principles of self-sufficiency, resilience and adaptation to a changing world, consistent with the ethos of both lifelong learning and sustainable development (Mongsawad, 2010). There are more than 23,000 such projects in rural villages across Thailand. Particularly relevant to the older Thai population is how this model has been strongly influenced by Thai Buddhist principles and the recognised value of traditional or local knowledge in rural contexts. In addition, such projects have also often provided new life skills and personal development opportunities for older persons and younger peers. The learning community's approach includes a focus on collaboration, problem-solving and 'thriving not just surviving' in changing times. This allows participating elders to take leadership roles and to contribute in a 'geragogic' way that includes 'collaborative peer' exchange as discussed below. In other words, as is the case with many community college 'continuing education' short courses, the roles of teachers and learners tend to be more flexible and ultimately interchangeable, as such projects and courses are guided by facilitators or mentors rather than teachers or trainers (Intarakumnerd, 2012).

The Guiding Role of the Silpakorn University U3A Centre

The Silpakorn University U3A centre is currently coordinating an initiative to promote the U3A banner in Thailand. It is currently in the process of creating a network of ten other local community groups that might also become U3A centres themselves. This is related to a strategy to enlist existing seniors' clubs and non-formal education centres in regional areas of the country, particularly through the promotion of education and cultural exchange. It also serves as basis for a related participatory research project about 'community mobilisation for promoting active ageing'. A key impetus of all this has been the assistance of Thomas Kuan who was a founder of one of the two Singapore U3A centres and who continues to promote the development and networking of U3A centres in the region.

Facilitated by the university's Department of Lifelong Education, the Silpakorn centre began as a seniors' club for older persons—active and retired—from various faculties across the university. It was developed to include courses and activities for local cultural and traditional knowledge exchange; for physical and mental health; recreation; and some language and life literacy (e.g. financial and internet/social media) courses. The emergence of the Silpakorn U3A centre provided a basis for going beyond the initial university origins to involve older persons from the local communities and other provinces in Thailand. The Silpakorn centre has been particularly keen to invite exchanges with local communities which are interested in sharing local expertise or knowledge in various cultural arts and forms of traditional knowledge. Such exchanges also include the kind of courses and activities regularly facilitated by Silpakorn U3A members. For health, they particularly promote courses in yoga, Tai Chi, meditation, health foods, exercises and basic self-care. More recreational activities include painting, dancing, singing, cycling and karaoke singing.

The importance of cultural exchanges to spreading the U3A network in the Thai context is exemplified by how the Silpakorn centre comprises of the following communities: Huey Duan Village in Nakhon Pathom Province, Ban-Rai Village in Ratchaburi Province and Nong-Sa-Rai Village in Kanchanaburi. Locals from both the Huey-Duan village and Ban-Rai villages have conducted courses or activities in cooking, handicraft, dancing and Buddhist practices. At the same time, other locals from Nong-Sa-Rai Village have similarly conducted activities of traditional dancing, singing, and folk music. In this way, Silpakorn U3A has been able to promote within its centres courses in ram klong-yao (long drum dance), ngan *baitong* (banana leaf craft) and *phuang ma hod* (Thai paper art) as well as coconut leaf craft, the art of folding lotus flowers, and traditional ceremony activities.

The Silpakorn U3A centre has therefore been active in exploring a relevant third age learning 'curriculum' for the Thai context. This has involved engagements with both the informal learning activities of local 'seniors clubs' and the geragogic-type courses for seniors conducted by some non-formal education centres. It links with both the Buddhist and local traditions of cultural activity and artefacts, encouraging the harnessing of experiential and traditional knowledge of Thai seniors. In addition to promoting this as collaborative 'peer sharing', the centre also organises visits by seniors to local schools and community groups, allowing them to share their knowledge and experience with younger generations and the wider society.

To facilitate networking among other U3A centres in local and surrounding area communities, the Silpakorn U3A centre adopts a model generally used by U3A centres in Singapore and other areas in the Asia-Pacific region. The Silpakorn U3A centre retains a supporting university context and recognises that local universities help promote the U3A concept in the Thai context. This is exemplified by the research project mentioned above which includes collaborations with three 'best practice' examples—namely the senior quality of life development centre in Nonthaburi (near Bangkok), the Berk-Prai sub-district seniors club in Ratchaburi province and the Thoem-Thong sub-district seniors college in Nan province in Northern Thailand.

Towards an 'Optimal' U3A Model of Seniors' Education and Learning?

In his earlier paper covering a history of the U3A movement, Formosa (2014: 42) expressed hope that 'the U3A movement will continue to be relevant to incoming cohorts of older adults by embracing a broader vision of learning'. In this respect, the Silpakorn University's initiative to support a local U3A model seems to represent a uniquely Thai effort to apply and/or develop a locally relevant U3A model. It might not just be considered in terms of how this model represents a relevant negotiation of the conventional U3A intersection between French and English models (that is, more academic/top-down vs. community-based/collaborative in approach). The model also exemplifies the wider cross-cultural challenge of developing a globally relevant and more 'optimal' model of later life, or seniors' lifelong education and learning. It represents an effort to harness both the advantage of links to a local university and the British model emphasis on local community hubs and wider network associations of 'collaborative peer' interactions and informal courses, or learning activities of interest. Whereas the latter emphasises a view of the 'lifelong learning' process as pertaining to health, the former is perhaps best exemplified by the 'lifelong education' policies of the European Union and their tendency to assume that seniors in ageing societies may need to continue working beyond normal retirement for sheer financial survival (Zarifis & Gravani, 2014). As Formosa argued, overcoming French-British polarities is a step in the right direction for the future of the U3A movement:

> Rather than entrenching the U3A experience in an absolutist vision–advocating either strict autonomy or complete integration with traditional universities - U3As have much to gain from seeking partnerships with tertiary educational sectors working within a similar ethos. Whilst partnerships in older adult learning do not have to be formally constituted and grand affairs, the benefits of collaborative approaches include that 'better information is available to help plan for learning, to deliver it in the best way, to promote engagement with it and to provide progression routes from it'…
>
> Formosa (2014): 58

Thus, this may also provide focus for exploring an 'optimal' model for older persons in terms of a more balanced integration of both lifelong education and learning as distinct if not complementary and convergent processes. Framed in terms of a larger study of non-formal education, which also makes reference to Thailand, Alan Rogers (2004) provided a relevant definition of 'informal lifelong education' which is particular useful here. This is especially in so far as Rogers' use of the term not only refers to a basic distinction between externally and/or socially 'structured' *education* and the individual cognitive 'process' of *learning*, but also informal, non-formal and formal modes of both as a continuum of sorts. Rogers thus describes informal lifelong education as the foundational aspect of a person's 'total lifetime learning' also including formal schooling as well as the ever-present possibility of informally learning from experience in life. Therefore, it makes sense for a globally relevant U3A model to both link to institutions of 'higher education and learning' and also frame local community contexts and processes of collaborative peer learning (that is, in some complementary sense harness aspects of both the French and English U3A models). And as discussed in the previous section, the emerging Silpakorn University U3A centre model appears to provide a basis for integrating and balancing social and psychological/cognitive modes of an education vs. learning continuum, harnessing the resources of both the local university and local community of seniors.

In this way also, such a model might more crucially (Formosa, 2014) use the 'geragogy' approach (alternatively referred to by some, as the 'eldergogy' approach) which is typically associated with the British U3A model. Whilst the term generally means a different teaching approach for seniors, it also has specific association with the 'collaborative peer learning' concept incorporating related notions of social networking or interaction and more informal notions of the teacher–student relationship and related roles (e.g. John, 1988). In contrast to the concept of *pedagogy*, which is synonymous with a teacher-centred perspective of formal education and learning (e.g. Merriam, 2001), the concept of *geragogy* has some overlaps with the use of the term *andragogy* in adult education. This is not just in terms of more collaborative and learner-centred approaches to teaching, but also in terms of the fact that life experience in general becomes a much more critical factor in the effective education and/or experiential learning of adults and seniors. Likewise, if extrinsic motivation is the dominant rationale of early life education, then it might be said that intrinsic motivation is key to effective adult education and the primary rationale for lifelong learning in both the third and fourth ages (Gorges & Kandler, 2012).

Formosa's (2012) suggestion of a critical geragogy (assumedly adapting notions of a 'critical pedagogy' in formal education) indicates a non-existent or under-developed 'reflective' aspect which might be framed for seniors in a way that is more accessible to cross-cultural and later-life contexts. Just as the related term presupposes a more 'active' rather than passive learner, so too the revolutionary insights of neuroplasticity (i.e. the ability of the brain to reorganise itself and continue developing by forming new neural connections even in old age and until the end) are challenging the traditional notion that older people lose the capacity to learn. In other words, an active and healthy mind is required for learning, and this can be optimised by active thinking and reflection. Perhaps the most exemplary mode of this in older adult and third age education is the function of 'life reviews'. As Haber (2006) suggested, this function should be seen as an important and even critical aspect of later-life learning, and not just as a therapeutic function, but also to have another chance in later life to celebrate positive experiences or achievements, and to share such achievements with peers and eventually also to come to terms with their own mortality.

In the related 'guided autobiography workshops' run by Thomas Kuan in Singapore (Richards & Kuan, 2017), which the Silpakorn U3A centre is keen to try in the local context, variations of this are useful in harnessing online social media (i.e. as collaborative peer interactions) to stay in touch with other seniors who may be isolated, provide digital family histories and records (including digitised photographs, videos and oral accounts) and/or involve reminiscing through the formal or informal writing of memoirs. If a fourth age version of this is especially associated with the function of 'gerotranscendence' (Tornstam, 2005), or the importance of Erikson's (1998) related 'lifecycle completion' function, then a 'third age' version is especially linked to the related challenges of retirement, family growing up and/or other either external or internal factors that may further compound and not just precipitate a mid-life or later-life existential crisis. The life review function is especially associated then with the internal function of how, as Gergen and Gergen (1997) have pointed out, every person has a unique life story—just as the sharing and not just telling of stories is a key aspect of collaborative peer learning for older learning.

The foundations of a more integrated and optimal informal curriculum for older learners typically underlie ad hoc and formal U3A learning programmes (Ratana-Ubol & Richards, 2016; Richards & Kuan, 2017). For instance, in addition to a common focus on activities on physical health, different educational provision for older adults in cross-cultural contexts also typically involves informal learning programmes on financial literacy and digital competency, and beyond this, similar courses and activities of social and cultural importance and topical or personal interest. In short, the life reviews function of such courses for seniors as digital storytelling, memoir writing and 'guided autobiography' (Birren & Cochran, 2001) may all assist to achieve a better balance of the physical, social and cognitive or even transcendent domains (that is, body, mind and spirit) of an exemplary and informal seniors' (especially 'third age') lifelong learning curriculum. As the Silpakorn U3A initiative has recognised, this is especially so in the cross-cultural contexts of a related 'tradition vs. modernity tension' (Cf. also Richards, 2013).

In addition to the exemplary modes of physical health, financial literacy and technological literacy, human mental health is tied to the cultural, reflective and intellectual activities, topics and domains that typically involve different foci of interest in different local and national contexts. Whilst the residual importance of traditional cultural activities and knowledge to Thai seniors was indicated above, the 'life review' also involves a *reflective* aspect which may represent an opportunity for personal development relevant to both traditional and modern contexts—including the wider Thai educational context where the 'authority of teachers' is still a fairly ubiquitous value that tends to discourage critical questioning and reflection (The Nation, 2015).

Conclusion

To the extent that it represents a move away from the more academic French or European model linked to established universities, some might consider that a limitation of the British U3A model is its characteristic informal 'amateurism' in terms of informal and non-formal activities and courses (Holt, 2012). This is not so much in terms of any comparison with a formal top-down curriculum or pedagogy, but rather in relation to a tendency to ignore or miss out on the benefits of both a more critical perspective and research basis for promoting the post-retirement living options for older persons in ageing societies (Johnson, 2015; Pauluski, 2016). There are cultural reasons why the Silpakorn University U3A centre recognised an opportunity to use its authority and position to help promote local U3A hubs and networks in Thailand with a similar 'peer collaboration' approach found in other Asia-Pacific countries. However, the additional advantage of this strategy is reflected in how these partnerships with other seniors' clubs, and non-formal education centres in Thailand were also able to be framed as a collaborative and participatory research collaboration focused on 'community mobilisation for promoting active ageing'.

Such strategies and related projects are valuable in influencing what Thai society and Thai media are currently recognising as increasingly important national policy. In short, such an approach may have some international and cross-cultural relevance in terms of a possible balance between French and British U3A models. It may help to ponder the challenge of balancing the strengths and weaknesses of an overall contrast between the European tendency for a perhaps somewhat prescriptive and top-down approach and the British model that emphasises the growing relevance of local 'grassroots' U3A centres, and thus research importance of seniors' lifelong learning in ageing societies in a fast-changing complex world. More modern societies may learn from the residual Thai perspective in relation to this challenge. As much as an integrated lifelong education framework may assist the physical and mental health of seniors in ageing societies, modern global societies should recognise they have much to learn from both traditional knowledge and from better harnessing the lived experiences of their seniors.

References

Birren, J., & Cochran, K. (2001). *Telling the stories of life through guided autobiography groups.* Baltimore, MA: The John Hopkins University Press.

Chaitrong, W. (2017). Risks grow for an ageing population. *The Nation.* http://www.nationmultimedia.com/detail/Economy/30333636v. Accessed March 10, 2018.

Charrurrangsri, A. (2012). *The development of the essential living skills for successful aging among Thai urban elders* (Unpublished Ph.D. Dissertation). Mahidol University, Thailand.

DW News. (2018). *As society ages Thailand sends seniors to school.* http://www.dw.com/en/as-society-ages-thailand-sends-seniors-to-school/av-43363678. Accessed March 15, 2018.

Erikson, E. (1998). *The lifecycle completed.* New York: WW Norton.

Formosa, M. (2012). Critical geragogy: Situating theory in practice. *Journal of Contemporary Educational Studies, 5,* 36–54.

Formosa, M. (2014). Four decades of Universities of the Third Age: Past, present, future. *Aging & Society, 34*(1), 42–66.

Gergen, K. J., & Gergen, M. M. (1997). Narratives of the self. In L. P. Hinchman & S. K. Hinchman (Eds.), *SUNY series in the philosophy of the social sciences. Memory, identity, community: The idea of narrative in the human sciences* (pp. 161–184). Albany, NY, US: State University of New York Press.

Gorges, J., & Kandler, C. (2012). Adults' learning motivation: Expectancy of success, value, and the role of affective memories. *Learning and Individual Differences, 22*(5), 610–617.

Haber, D. (2006). Life review: Implementation, theory, research and therapy. *International Journal of Ageing and Human Development, 63*(2), 151–171.

Holt, R. (2012). Amateurism and its interpretation: The social origins of British sport. *Innovation: The European Journal of Social Science Research, 5*(4), 19–31.

Intarakumnerd, T. (2012). The development of the community college system in Thailand. In A. Wiseman, A. Chase-Mayoral, T. Janis, & A. Sachdev (Eds.), *Community colleges worldwide: Investigating the global phenomenon* (pp. 375–383). Bingley, UK: Emerald Group Publishing Limited.

Jitapunkul, S., & Wivatvanit, S. (2009). National policies and programs for the aging population in Thailand. *Ageing International, 33*(1–4), 62–74.

John, M. (1988). *Geragogy: A theory for teaching the elderly.* Binghampton: The Haworth Press.

Johnson, S. (2015). How can older people play a bigger role in society? *The Guardian.* https://www.theguardian.com/society/2015/mar/30/how-why-older-people-valued-knowledge-experience. Accessed March 20, 2018.

Jungck, S., & Kajornsin, B. (2003). Thai Wisdom and Globalization: Negotiating the global and local in Thailand's national education reform. In K. Anderson-Levitt (Ed.), *Local meaning, global schooling* (pp. 27–49). Basingstoke, UK: Palgrave Macmillan.

Merriam, S. (2001). Andragogy and self-directed learning: Pillars of adult learning. In B. Merriam (Ed.), *The new update on adult learning theory* (pp. 3–13). Hoboken, NJ: Jossey-Bass.

Mongsawad, P. (2010). The philosophy of the sufficiency economy: A contribution to the theory of development. *Asia-Pacific Development Journal, 17*(1), 123–143.

Paluski, J. (2016). Facing the challenges of an ageing society. In H. Kendig, P. McDonald, & J. Piggot (Eds.), *Population ageing and Australia's future.* Canberra, Australia: ANU Press.

Ratana-Ubol, A., Charungkaittikul, S., & Sajjasophon, R. (2012). *The scenario of U3A for lifelong learning of Thai aging.* Bangkok: Chulalongkorn University Publication.

Ratana-Ubol, A., & Richards, C. (2016). Third age learning: adapting the idea to a Thailand context of lifelong learning. *International Journal of Lifelong Education, 35*(1), 86–101.

Richards, C. (2013). What can China and the West still learn from the other in new times? Towards a social psychology of comparative knowledge systems. In J. Zhang, Y. Yang, L. Lui, & M. Zhou (Eds.), *Towards social harmony: A new mission of Asian social psychology* (pp. 94–119). Beijing: Educational Science Publishing House.

Richards, C., & Kuan, T. (2017). Life reviews and seniors lifelong learning for lifecycle completion, In C. Richards, & Charungkaittikul, S. (Eds.), *The eight pillars of lifelong education*. Chulabooks: Thailand studies.

Rogers, A. (2004). Looking again at non-formal and informal education. Towards a new paradigm. *The Encyclopedia of Informal Education*. www.infed.org/biblio/non_formal_paradigm.htm. Accessed April 5, 2018.

Thailand Constitutional Court. (2007). *Constitution of the Kingdom of Thailand B.E.2550*. Bangkok: Bureau of Printing Service.

The Nation. (2015). Thailand schools should focus on critical thinking, *Straits Times*. https://www.straitstimes.com/asia/se-asia/thailand-schools-should-focus-on-critical-thinking-the-nation. Accessed March 12, 2018.

Tornstam, L. (2005). *Gerotranscendence: A developmental theory of positive ageing*. New York: Springer.

United Nations Educational, Scientific and Cultural Organization. (2016). *Monitoring results for Thailand, 3rd Global Report on Adult Learning and Education (GRALE)*. http://uil.unesco.org/system/files/thailand.pdf. Accessed March 10, 2018.

World Bank. (2016). *Thailand Economic Monitor—June 2016: Aging society and economy*. http://www.worldbank.org/en/country/thailand/publication/thailand-economic-monitor-june-2016-aging-society-and-economy. Accessed March 5, 2018.

Zarafis, G. & Gravani, M. (Eds). (2014). *Challenging the "European area of lifelong learning": A critical response*. Dordrecht: Springer.

Cameron Richards is a semi-retired Professor with extensive experience in the Asia-Pacific region. He has a multi-disciplinary background which includes specialisations in sustainability studies, educational policy research and lifelong learning. In addition to his regular sustainability projects consultancy work, he has held past lecturer (and currently, adjunct) positions at universities in Australia, Singapore, Hong Kong, Malaysia and Thailand. In recent years, a central focus of his work has been on developing an integrated lifecycle model of lifelong education and learning—one which has particular application to the area of 'later life learning'. In addition to his 2017 book *The eight pillars of lifelong education*, his recent papers include publications in the *International Journal of Lifelong Education*, *Journal of Adult and Continuing Education*, and the *Journal Research on Adult Education*.

Jittra Makaphol is former Head of the Silpakorn University's Lifelong Education Department within the same university's Faculty of Education. Her research interests include active ageing, local wisdom and teaching and learning of older persons. She has been honoured as the Outstanding Researcher of the Silpakorn University's Education Faculty in 2016 and 2018. Prior to joining Silpakorn University, she had served as an educator of the Thailand Science Centre and a senior supervisor of the Office of Non-formal and Informal Education in the Thailand Ministry of Education. She has more than twenty years of experience in the areas of adult education and lifelong learning and has earned a Bachelor of Science in Physics from Chulalongkorn University, and a Ph.D. in Lifelong Education and Human Development from Silpakorn University.

Thomas Kuan is the Founder of the U3A centre in Singapore and the President of East Asia Federation for Adult Education (EAFAE), 2017–2019. He is currently the Treasurer of PIMA, the PASCAL International Member Association. He currently sits on the editorial team for the virtual U3A Signpost e-Newsletter. He holds an MA in education, training and development (Hull), and diplomas in management consultancy, human resources and industrial engineering. He is Fellow of the Beta Phi International Literary Society.

Part IV
Coda

Chapter 21
Concluding Remarks and Reflections

Marvin Formosa

This final chapter provides an opportunity to reflect on the significance of the travails of the University of the Third Age (U3A) movement which, since its modest inception in 1973, has certainly exceeded all expectations. It does not strive to tie loose ends in a steadfast manner. Rather, the chapter serves as a beacon for stimulating further theoretical and empirical research on third age learning and the U3A in particular, by tracing the similarities and divergences of U3As across European and Asian-Pacific contexts; the benefits of participation in U3As on learners' quality of life; and their social and psychological well-being. However, this chapter also points out that the U3A's track record in the democratisation of late-life learning—especially in terms of social class, gender, disability and ethnic biases—leaves much room for improvement, to the extents that one can conclude that many U3As are reinforcing a degree of inequality amongst older persons in general but especially amongst the most vulnerable sectors of the ageing population. The chapter's final section highlights the need for the U3A movement to renew itself so as to remain relevant to incoming groups of older persons whose generational habitus is certainly distinct from that of preceding cohorts—to whom the U3As are generally targeted—by including a number of policy recommendations for the future.

Similarities and Differences

All the chapters within this edited text highlighted a relatively common philosophy amongst U3As in the various countries and two continental locations under focus. U3As are generally united in their efforts to provide learning opportunities for older adults, as well as to increase the visibility of older generations whose presence and worth tend to be universally undervalued and discounted. U3As also remind civil

M. Formosa (✉)
University of Malta, Msida, Malta
e-mail: marvin.formosa@um.edu.mt

© Springer Nature Switzerland AG 2019
M. Formosa (ed.), *The University of the Third Age and Active Ageing*, International Perspectives on Aging 23, https://doi.org/10.1007/978-3-030-21515-6_21

society and political parties of the actual meaning of the term '*lifelong* learning', whilst providing a niche for a category of citizens who are generally left out in the cold as far as learning opportunities are concerned, since governments tend to cling to traditional models of education that promise higher levels of production, profitability and employability. Many authors stressed how U3As in their countries contribute significantly to active ageing, as participation in learning programmes encourages older people to move out of their homes to join peers in physical, cognitive, expressive, creative and social accomplishments. To take a feather out of Iceland's cap, the mantra of U3As in this country is that—irrespective of one's age—it is never too late for older persons to rethink their situation in life, embark on new missions, fulfil ambitions and act upon dreams and desires. Regardless of whether U3As are fiercely autonomous centres, embedded within traditional universities or instigated by the government, all U3As strive to enable older people to engage in the sharing of knowledge, skills, interests and experience; determine the benefits and enjoyment to be gained, and the new horizons to be discovered through late-life learning; and ultimately, to applaud the capabilities, potential and social value of older people. In their attempt to reach out to a wider and more numerous audience, U3As endeavour to overcome any defensiveness on behalf of older persons about participating in late-life learning, usually driven by the old adage that 'you cannot teach an old dog new tricks'. Moreover, U3As are ready to collaborate with educational institutions to undertake research into ageing and the position of older people in society. Again, one finds Iceland's U3A in Reykjavik to be at the forefront of such initiatives, as the BALL project received the Erasmus + Award for Excellence in the field of adult education, with its results having been published in the book *Towards a dynamic third age* (Guðjónsdóttir et al., 2016) which includes guidelines and recommendations on how to best prepare for the third age. Indeed, the majority of U3As based in Europe take part in projects funded by the European Commission such as Erasmus +, Tempus, and until its termination, Grundtvig. For example, the Maltese U3A took part in a Tempus project entitled Centre for Third Age Education (n.d.) with 14 other U3As in countries like Italy, Slovakia, Russian Federation, UK, Spain, Azerbaijan and Ukraine. The objectives of this project included raising awareness of third age learning in such countries, to establish new U3As in all these countries, and to coordinate intergenerational learning projects. This Tempus project succeeded in creating third age education centres in all the partner countries, purchasing equipment for the running of these centres, developing study programmes and curricula for each centre, as well as providing training to 180 teachers and facilitators in the field of geragogy. The project culminated in an international conference on third age learning in Genoa which also included the launch of the publication *Active ageing: From wisdom to lifelong learning* (Amoretti, Spulber, & Varani, 2017) which collected and presented the different experiences that separate countries face in promoting and implementing third age learning, whilst documenting the best practices implemented in the various sociocultural contexts that characterised the partner countries. The International Association of Universities of the Third Age (n.d.) also engaged in a number of research projects on themes ranging from senior tourism to art appreciation in later life, to equal learning opportunities across the life course.

However, the chapters in this edited text also uncovered a long-standing schism in both the *modus operandi* and *modus operatum* of U3As. On the one hand, the affixation of Francophone-inspired U3As to traditional universities—as is the case in France, Germany, Lebanon, Spain and Malta—brings about both potential advantages and risks. Although such U3As are generally successful in accessing funds, retaining nominal enrolment fees (as they make free use of university resources) and implementing any study programmes that strike the members' fancy due to the large pool of teaching personnel at the university, classes tend to take place on university campuses which tend to be far away from residential centres, and thus, relatively inaccessible. Moreover, Francophone U3As run the risk of being characterised by a lack of agency, as the academic body has the last say on most matters and total decision-making power on aspects related to learning. On the other hand, the self-help approach favoured by U3As following the British model—as in the UK, Australia and New Zealand—opens the possibilities of nominal membership fees; accessible classes run in community halls, libraries and private homes; flexible curricula and teaching styles; wide course varieties that range from the highly academic to arts and crafts; no academic constraints such as entrance requirements or examinations; and the opportunity to mix with alert like-minded people who enjoy doing new things. Nonetheless, drawbacks are also present. First, the British U3A movement has become the victim of its own success in that its triumph in attracting more learners gave rise to issues with locating suitable venues, and enough volunteers to run courses of study and interest groups. Second, Laslett's (1989: 22) criterion that 'no support from the funds of local or central government shall be expected or sought' is debatable. Whilst such a stance renders such U3As at risk from being marginalised and disengaged from potential valuable allies, Freire's (1972) dictum that no learning sphere can ever be politically neutral makes such a claim of political impartiality highly dubious. After all, one of the founders of the British U3A movement was categorical that the U3A in Britain was, from the outset, hinged upon the ideals of 'utopian socialism' (Midwinter, 2007). Finally, whilst it is possible that the autonomy of teaching and learning in the U3As will be impaired if facilitators participate in uniform and standard training courses, it is equally conceivable that they may not have the ability to teach or communicate well. Third age learning diverges from the basic pedagogical and andragogical principles and is embedded in geragogical philosophies and practices (Formosa, 2002; Gómez, 2016). Of course, there are U3As which do not fit either of these two models. Those of special interest include U3As in Russia, China and Taiwan, which emerged neither as part of traditional universities, nor as self-help organisations, but were directly established, managed and financed by government authorities and ministries as part of the countries' official state policy. Although a passionate debate on the pros and cons of such a state of affairs is understandable, the fact that there is no unitary manner as to how U3As have been launched and preserved contributes further to the richness of the third age learning movement. Indeed, despite the idiosyncrasies characterising the various formats of U3As, the evidence paints a picture of a dynamic, flexible, accessible learning movement which is meeting the needs and interests of a rapidly growing number of older adults without the necessity of binding script. Whilst entertainment activities such as bingo may well remain a favourite option for some elders, it is establishments such as U3As which will be increasingly sought in the coming years considering that each incoming generation holds higher levels of educational attainment and qualifications than the preceding one.

Universities of the Third Age as Vehicles for Active Ageing

As active ageing refers to 'the process of optimizing opportunities for health, participation and security in order to enhance quality of life as people age'—whereby the term 'active' refers to 'continuing participation in social, economic, cultural, spiritual and civic affairs, not just the ability to be physically active or to participate in the labour force' (World Health Organization, 2002: 12)—it is clear from all chapters in this manuscript that U3As in the European and Asian-Pacific contexts do contribute to the improvement of the levels of active ageing in their respective countries. There are unmistakable health benefits from participating and attending learning programmes at U3As, benefits that can be split into physical and mental well-being. The U3As' unique blend of learning and social activities places an unorthodox emphasis on autonomy and participation, and hence, countering the 'decline and loss paradigm' commonly associated with increasing chronological age. Since the beneficial impacts of joining a U3A traverse both health and social boundaries, U3As accentuate the need for a departure from notions of active ageing in purely productive terms, to one that embraces mental and physical well-being, and social participation. Participation in U3A centres provides members with a renewed focus in their personal lives, which strengthens their mental well-being by bolstering their sense of purpose, self-confidence and self-worth. This is especially valuable to older persons who have not yet come to terms with retirement, who are still experiencing a sense of bereavement and social alienation following the loss of a working day's structure, and lifelong colleagues and friends. As a recent study reported,

> Involvement was described as having focus, and a purpose for the day which ensured mental wellbeing…Learning subjects such as languages and music were described as being beneficial to memory. The opportunity to face new challenges was perceived as having a positive effect on health and wellbeing.
>
> Third Age Trust (2018): 13

The above is far from surprising since a study on the relationship between emotional well-being (operationalised as autonomy, personal growth, control, positive relationships with others, purpose, personal acceptance and generativity) and participation in the São Paulo U3A in Brazil concluded that 'the students [sic] who had been longer on the program run by the institute studied, exhibited higher levels of subjective and psychological well-being…where the satisfaction and benefits gained [from learning] extend into other areas of life' (Ordonez, Lima-Silva, & Cachioni, 2011: 224). The chapters herein also point out how for many U3A members, friendship and feeling connected to others were equally beneficial, especially when one considers the difficulties in forming new friendships following work exit and widow-hood, especially for older men and women, respectively. U3As have much potential in permeating members with improved levels of bridging and bonding types of social capital and may arise as an arena of social cohesion by providing mutual support in times of life crisis or difficult life transitions such as in the case of bereavement, sickness or moving house. In the UK, it was reported that as

> …most U3A activities take place during the day and on weekdays, participants reflected that it provided a good replacement for the previous work time…joining the U3A was a planned positive step in retirement offering the opportunity to meet new people who were different from those in the workplace…
>
> Third Age Trust (2018): 13

This also means that U3As hold a strategic role in mitigating against loneliness and isolation, as many members offer continuous support and presence to each other on a daily basis. The arising sense of social cohesion is particularly evident when one realises that the U3A movement is an important part of the lifelong education system, a key segment of building a learning and intergenerational society, one of the basic public services enjoyed by civil society, and hence, acting as a channel towards active citizenship in later life. U3As are also key catalysts for active ageing as they succeed in affecting important and far-reaching life changes to members. To cite again the recent study published by the Third Age Trust,

> Many participants spoke about people who had been bereaved whilst being part of the U3A. The membership offered support. A phrase used was that the U3A "folded around" the bereaved member, so they are safe in the knowledge that they are part of a group that cares about them. The U3A remains a focal point for those trying to redefine their lives and sometimes themselves in this situation. Several times, the phrase "it saved my life" was used…The same 'group care' approach was also evident when sickness was discussed.
>
> Third Age Trust (2018): 13

Finally, the real thrust in favour of active ageing enacted by U3As is witnessed by the centres' engagement with communities. The Third Age Trust (2018) outlined how U3As in England supported local community halls, libraries, environmental campaigns, art centres, churches, village fetes, dementia support days, as well as music festivals. Another research study reporting on the Choir of the U3A Hawthorn, Melbourne, Australia, and the benefits perceived by members undertaking this active music engagement in non-competitive choral singing found that

> …participants who are all members of the Choir of the Hawthorn U3A perceive membership of the group to be significant in their lives. Such programmes as the choir that is the focus of this study offer a way for seniors to be involved in an activity that engages and maintains well-being through forming and sustaining relationships within a community singing in a community choir can provide…musical and social experiences for a group of older people active in society. Such physical, mental and social engagement assists older people remain independent and active in the community; thus, delaying the need for residential care.
>
> Joseph and Southcott (2015): 345

Throughout the preceding chapters, this community engagement on behalf of U3As was especially evident in the Australian and Thai contexts. For instance, Lamb underlined how over time many U3As in Australia engaged with the wider community, by developing relationships with local government councils, local newspapers, radio and television stations, and other older age group organisations in their localities. Moreover, several U3As in Australia have taken some of their activities into retirement villages for the benefit of those residents in care units, aiming to outreach to older persons who are generally not typical members in U3As. Echoing Joseph and Southcott's (2015) sentiments, she concluded that U3A choirs often visit such villages to provide musical entertainment where residents can join in community singalongs. As access to participation in music, dance and art is of particular value to older people, especially those suffering from dementia, and since these activities stimulate feelings of happiness and well-being and open up new channels of communication, the Australian U3A's efforts in favour of community engagement are resulting in direct positive impacts on the quality of life and emotional well-being of older persons residing in rural and pastoral parts of the country. As Richards and colleagues outlined in a previous chapter, in Thailand, the importance of cultural exchanges to spreading the U3A network is exemplified by how the Silpakorn U3A is able to reach a number of different communities, so that locals from separate villages participate in learning programmes or meaningful activities such as cooking, handicraft, dancing, Buddhist practices, singing and folk music. In this way, the Silpakorn U3A was successful in incorporating a number of surrounding communities in both the construction and implementation of its learning ethos and vision. Since people do not age actively in a social vacuum but do so by continuing their participation in social and cultural events, the U3As hold much potential in becoming efficient and effective vehicles that enable citizens to reach higher levels of successful and productive ageing.

Quandaries and Lacunae

Whilst the U3As' beneficial impacts can never be overestimated, it is nevertheless clear from the chapters in this volume that the movement is entrenched within a Parsonian vision of the world—namely a normative social system in which superficially divided groups are united by shared values (Parsons, 1951). Although U3A members do assist and support each other in times of difficulties, problems are always approached—to use Wright Mills' (1959) terms—as 'private troubles' rather than 'public issues'. Moreover, as U3As refrain from engaging in age-related advocacy, they create a 'neutral' space for the practice of older adult learning, which ultimately functions as an arena for the safeguarding and protection of a third age lifestyle. This, in turn, compels the U3A movement to practise 'social closure'—namely an engagement in drawing boundaries, constructing identities and building communities in order to monopolise scarce resources for one's own group, thereby excluding others from using them (Weber, 1978). Although this is mostly unintentional, practices of social closure function to dissuade subaltern older persons from enrolling and participating in U3A learning programmes and social events. The subsequent subsections highlight the role of U3As in reproducing unequal relations in later life with special emphasis on elitism, gender, racial/ethnic bias, third ageism and internal ageism.

Social class. Researchers have long commented how 'threatened…by elitism and pastime activism, U3As might indulge in narcissism and escapism and miss altogether the highest vocation they should respond to' (Philibert, 1984: 57), and how U3As 'pandered to the cultural pretentious of an aged bourgeoisie who had already learned to play the system' (Morris, 1984: 136). The situation remained unchanged, and typical U3A members continue to hold higher-than-average levels of socio-economic status. On the one hand, for middle-class older persons, joining U3As means going back to an arena in which they feel confident and self-assured of its development and outcome. Moreover, for middle-class older persons U3A membership becomes a strategy of 'distinction' (Bourdieu, 1984) in the way that coffee-table books and paintings are used to impress friends and other social viewers. On the other hand, working-class elders are apprehensive to join an organisation with such a 'heavy' class baggage in its title. Moreover, the liberal-arts curriculum promoted by most U3As is perceived as alien by working-class older persons, who tend to experience 'at-risk-poverty' lifestyles and are more interested in vocational knowledge and practices.

Gender. U3As include gender biases that work against the interests of both men and women. On the one hand, although U3As include a higher percentage of female members, studies showed that they are generally anchored in gender expectations about women's traditional roles. For instance, whilst male learners are more likely to dominate any discussion even when in the minority, U3As also tend to be characterised by a 'masculinist' discourse where women are silenced and made passive through their invisibility (Formosa, 2005). On the other hand, men do not enjoy any preferential treatment since most curricula tend to reflect the interests of females (Formosa, Chetcuti Galea, & Farrugia Bonello, 2014; Formosa, Fragoso, Jelenc Krašovec, & Tambaum, 2014a, b). For instance, health promotion courses are generally delivered by female tutors with a bias towards women-related health issues such as weight loss and osteoporosis. U3As are yet to latch on to the successes of the Men's Shed movement which notes how men may not embrace conventional formal learning contexts, yet participate enthusiastically in community organisations—such as sporting and hobby clubs (Golding, 2015)—which can be counteracted by setting male-friendly curricula that attract higher membership rates of older men. It is positive to note that, as illustrated in a previous chapter, the Maltese U3A has taken some mitigating steps and action towards increasing the male membership.

Racial/Ethnic bias. U3A membership, even in large multicultural cities, tends to lack ethnic minorities in its membership body. Findsen and colleagues (2017) argued that this may arise due to the fact that the projected ethos of U3As mirrors the values of the dominant groups in society, so ethnic minorities feel that they do not have the necessary 'cultural capital' to participate. This critical perspective was recently substantiated by Ratana-Ubol and Richards (2016: 92) who pointed out that U3As were ultimately created in—and for—a Western context so that there might be 'some difficulty with or resistance to efforts to link and frame and extend…the U3A concept specifically [as] an international movement'. The authors proposed that for the Thailand context (and for any other that finds similar resistances or difficulties in framing senior citizen lifelong learning in terms of the U3A concept)

> ...the term *third age learning* should be preferred...adding a cross-cultural and international dimension to this. Third age learning...helps to cut across the *tradition-modernity divide* and related 'divides' or issues without losing the essential message and attraction of an increasingly significant overall concept - a concept too important to lose because of some cultural issues and semantic dilemmas.
>
> Ratana-Ubol and Richards (2016): 96 - italics in original

Third ageism. Although Lamb in this volume noted how Australian U3As conduct learning sessions in retirement villages, it is ostensible that U3A membership does not generally include older persons with mobility and cognitive difficulties. U3As hinge their outreach work upon Laslett's (1989: 4) definition of the 'third age' as a 'period of personal fulfilment, following the second stage of independence, maturity, responsibility, earning, and saving, and preceding the fourth age of final dependence, decrepitude and death'. Henceforth, sparse effort is made to outreach those persons in the 'fourth age'—namely who are homebound or living in residential care, and thus, unable to reach classroom settings. As Indeed, Robbins-Ruszkowski (2017: 2017–8) articulated, 'despite attending UTAs in two cities over a period of four years, I never saw anyone in a wheelchair. Very few people used a cane. All meetings were held in university spaces across the cities, to which people arrived by bus and tram'. This is problematic because both young-old persons with early-onset complications from strokes, diabetes and neurological diseases and old-old persons with mobility and mental challenges may still harbour learning needs and interests (Formosa & Cassar, forthcoming). Although one notes some efforts to outreach homebound older persons through online strategies, U3As generally overlook how learning opportunities can positively impact the personal and social development of frail older people, as well as older persons with dementia.

Internal ageism. Many U3A members use the learning prospects and the arising camaraderie at centres as strategies to resist, and some even to deny their transition into ageing lives. Certainly, U3As may be seen as vehicles through which middle-class and able-bodied older persons mitigate against what Brown (n.d.) terms as 'oldering'—namely a type of labelling, a specific form of Foucault's (1982) objectification, a process whereby older persons make themselves subjects, accepting inferior status and ageist discrimination. Whilst such a stance may seem a positive one from an advocacy point of view, it is also somewhat detrimental. Reminiscent of Wilinska's (2012) research, most U3As in European and Asian-Pacific milieus strive to reject the idea of 'old age' rather than resisting ageist discourses, rarely outreaching older peers who have little in common with typical U3A members, and hence, playing only a minor role, if any, in changing the social circumstances of older citizens. In Karpf's (2014: 42) words, U3As are 'just gerontophobia in its latest carnation', just like when the 'magazine articles that tell you "How to Look Fabulous At Fifty"...they encourage you to deal with prejudice not by challenging it but by trying not to look old'. Indeed, 'inasmuch as U3As promote an increasingly valued cultural model of aging, it is necessary to highlight the hidden ways that they can reproduce existing inequalities' (Robbins-Ruszkowski, 2017: 2019).

Renewing Universities of the Third Age

As the U3A movement surpasses its 45th year of operation, its key challenge is two-pronged. The first is for U3As to remain relevant and attuned to the life-world of present and incoming older cohorts. The U3A movement generally overlooks how incoming older cohorts are characterised by diverse generational dispositions when compared to those held by older adults during the early 1970s and 1980s when Vellas and Laslett conceptualised and launched their Francophone and Anglophone U3A versions. Despite its promise, the notions of 'activity theory' and 'active ageing' are fast becoming inadequate to capture the complexity of older persons' lives in the face of changing family dynamics, increasing individualisation, the all-encompassing online world, and a burgeoning silver industry (Moody & Sasser, 2014). The U3A movement includes no impassable shield against what is termed as structural lag—namely a failing on behalf of structural arrangements to meet or be relevant to the needs of a large proportion of the clientele they are supposedly serving (Riley & Riley, 1994). In their uncritical espousal of the benefits of active ageing and older adult learning, U3As are running the risk of ignoring how daily lives in the third age are changing over time. Indeed, U3As must reflect on how the mantra of 'keeping active' has turned into a moral imperative, and need to heed Katz's (2000) advice to unpack a further notion of learning in relation to later life. This may mean going beyond the safe waters of 'individualised' learning for personal satisfaction, to build bridges with the realms of productive, positive and spiritual ageing. Older persons are becoming less amenable to be treated as empty receptacles for the deposit of knowledge, and strive to make significant economic contributions to society, especially in terms of employment, volunteering, caregiving and formal accredited education (Foster & Walker, 2015). The fact that a record number of older persons are choosing to remain in the workforce for much longer than previously indicates the need to understand older lives differently. At the same time, retirement is no longer merely the onset of social and cognitive finitude, and ageing is now perceived to offer the chance of making sense of persons' lives. As Cohen (2005) argued, retirement can represent a liberation phase, when people feel a desire to experiment, innovate and skirt around social conventions to explore new paths to creativity. Finally, older persons are rebutting the old adage that no reference to ageing as such, or increasing birthdays, should be made in elder learning. Scholars and learning facilitators such as Moody (1990) and Russell (2008) have explored the possibilities of 'transcendental' learning through which older adults reflect on their lives and repair relationships, seek to understand their role in the world and acknowledge their impending death. The extent to which U3As must undergo an inclusive and extensive renewal process becomes even more warranted when one considers the vast range of older adult lifestyles in late modern societies. These range from (i) golden years where relaxing time remains the dominant preoccupation; (ii) neo-golden years in which the emphasis is on pursuing self-development; (iii) portfolio life where the individual seeks a balance between work, family responsibilities, leisure pursuits and volunteering; (iv) second career in a new employment arena; and (v) extension of a midlife career where an individual continues his/her existing career as long as possible (Kidahashi & Manheimer, 2009). In view of such a wide diversity of lifestyles, Withnall (2010) argued persuasively that the concept of lifelong learning really has no practical future on account of the need to develop any all-embracing framework within which dissimilar challenges could be addressed. Instead, she believed that

> …it is more important to think in terms of 'longlife learning', a concept that would encompass a whole range of economic, democratic, personal and other concerns throughout the life course whilst highlighting the enormous impact of demographic trends both now and in the future. Learning would come to be recognised in all its different forms at all ages and would contribute not just to economic progress and social inclusiveness but to people's desires for personal development and creativity as they grow older. Such learning would not necessarily be linear or even cumulative in the sense of building up a clearly defined bank of knowledge and skills, but it would allow for a personal exploration that makes sense at an individual level and would be enduring and connective…
>
> Withnall (2012): 663

The second challenge refers to the required practical and pragmatic strategies to steer the U3A ethos away from one of 'lifelong learning' and towards a 'longlife learning' one. This goal necessitates four concurrent strategies.

Overcoming French-British polarities. Contrary to what is generally assumed, learning as an 'end-in-itself' and learning for educational accreditation are not necessarily incompatible and may even be complimentary. Whilst many older adults emboldened by their U3A experience go on to a tertiary educational programme, numerous older adults in tertiary education enrol in U3As (Formosa, 2014). Rather than entrenching the U3A experience in an absolutist vision—advocating either strict autonomy or complete integration with traditional universities—there is much potential in seeking partnerships with tertiary educational sectors, government agencies, local authorities and non-governmental organisations. The chapters in this volume noted many such partnerships which resulted in a range of benefits, from rent-free premises which allowed low-cost fees to easy access for invited speakers and members to intergenerational solidarity. However, the possibility that U3As become absorbed as part of the silver industry, along with culture and travel, and an opposition to a 'commodified', needs to be resisted as this can blur the line between humanist and emancipatory learning with commercially run educational-travel enterprises (Moody, 2004).

Quality of learning, instruction and curricula. The precise contribution of learning in U3As to an empowerment and emancipatory agenda remains elusive. One must ask whether learners at U3A centres are too docile, too passive, as though listening alone were enough. This warrants a learning environment in which 'learners who are able to take control and direct learning…who, in their daily lives, know how to put into practice learning they have undertaken' (Gladdish, 2010: 15). The quality of instruction is also to be put under scrutiny. Facilitators should enable older adults to foster the control that they may be consciously or subconsciously lacking through encouragement to take responsibility for their learning by choosing those methods and resources by which they want to learn. Learning in later life thrives on collaboration and partnership, characterised by 'co-operative work' between tutors and learners. Moreover, U3As must not assume that older learners continue living in some bygone world. Rather, e-learning has become increasing popular in later life as it offers the opportunity for older learners to access information and communicate with others when and if they want to. For U3As to continue being relevant to contemporary elders, centres must make more effort to embed their learning strategies in the web 2.0 revolution, which contrary to its predecessor uses interactive tools—ranging from Blogs, Wikis, Podcasts, online journals, to virtual picture databases—to offer limitless possibilities for an interactive, empowering and participatory form of older adult learning. The curriculum at U3As should be as bold and original as possible, negotiated with and even determined by the varied interests of learners. In this respect, there is a real urgency for U3As to include non-liberal and health-related areas of learning such as financial literacy and caregiving, as well as environmental, botanical and zoological studies.

Social inclusion. U3As include a relative absence of older men, older adults living in rural areas, older persons from working-class backgrounds, ethnic minorities and peers with mobility and cognitive challenges. Underlying reasons for such non-participation generally range from a lack of information, situational barriers such as caregiving duties, institutional barriers such as the middle-class and ethnocentric atmosphere of learning centres, and dispositional barriers such as negative attitudes towards learning in general (Formosa, 2016). Hence, special targeting and outreach strategies, bottom-up consultation and negotiation, and the usage of non-participants' environments, issues and concerns for programme development are warranted if U3As are to reach higher and more equitable levels of social inclusion. Distinct strategies may include providing adequate transport facilities to and back from the learning centres, and holding sessions in learners' residences, community halls, sports venues and residential long-term care facilities. Incorporating Men's Sheds as part of U3As that offer courses such as restoring furniture, woodworking, mechanics and other material skills will also serve to attract higher numbers of older men to U3As.

To conclude this chapter and book, one can never overemphasise the need for further research that continues mapping the national travails of the international U3A movement. Space limitations meant that it was not possible to include chapters on a number of established U3As in Europe and Asia—such as those in Belgium, Finland, Switzerland, the Netherlands, Portugal, Luxembourg, Greece, Czech Republic, Slovakia, Norway, India, Japan, Nepal and Singapore, amongst others. Attention must also be bestowed on other continents, especially South America, where the U3A movement is much alive and kicking. Space limitations also meant that the contemporary interface between U3As and novel communication technologies was not tackled in this text. Compared with the humungous nature of the U3A movement, and extensive work performed by both volunteers and salaried workers to keep U3As afloat, it is unfortunate that academic deliberations on this movement remain relatively sparse. It is sincerely hoped that this publication serves as a catalyst for the planning and carrying out of further theoretical and empirical research work on the U3A movement.

References

Amoretti, G., Spulber, D., & Varani, N. (Eds.). (2017). *Active ageing: From wisdom to lifelong learning*. Rome: Carocci editore.

Bourdieu, P. (1984). *Distinction: A social critique of the judgement of taste*. London: Routledge.

Brown, M. (n.d). *Learning in later life: Oldering or empowerment? A third-age researcher's interpretation of some voices of third-age learners*. Xeroxed copy.

Centre for Third Age Education. (n.d.). *Centre for Third Age education*. http://www.tempus-ctae. eu/. Accessed February 12, 2019.

Cohen, G. D. (2005). *The mature mind: The positive power of the ageing brain*. New York: Basic Books.

Findsen, B., Golding, B., Jekenc Krasovec, S., & Schmidt-Hertha, B. (2017). Ma te ora ka mohio/'Through life there is learning'. *Australian Journal of Adult Learning, 57*(3), 509–526.

Formosa, M. (2002). Critical gerogogy: Developing practical possibilities for critical educational gerontology. *Education and Ageing, 17*(1), 73–86.

Formosa, M. (2005). Feminism and critical educational gerontology: An agenda for good practice. *Ageing International, 30*(4), 396–411.

Formosa, M. (2014). Four decades of Universities of the Third Age: Past, present, and future. *Ageing & Society, 34*(1), 42–66.

Formosa, M. (2016). Malta. In B. Findsen & M. Formosa (Eds.), *International perspectives on older adult education: Research, policies, practices* (pp. 261–272). Cham, Switzerland: Springer.

Formosa, M., & Cassar, P. (forthcoming). Visual art dialogues with older people in long-term care facilities: An action research study. *International Journal of Ageing and Education*.

Formosa, M., Chetcuti Galea, R., & Farrugia Bonello, M. (2014a). Older men learning through religious and political affiliations: Case studies from Malta. *Androgogic Perspectives, 20*(3), 57–69.

Formosa, M., Fragoso, A., & Jelenc Kraŝovec, S. (2014b). Older men as learners in the community: Theoretical issues. In S. Jelenc Kraŝovec & M. Radovan (Eds.), *Older men learning in the community: European snapshots* (pp. 15–28). Ljubljana: Ljubljana University Press.

Formosa, M., Fragoso, A., Jelenc Kraŝovec, S., & Tambaum, T. (2014a). Introduction. In S. Jelenc Kraŝovec & M. Radovan (Eds.), *Older men learning in the community: European snapshots* (pp. 9–13). Ljubljana: Ljubljana University Press.

Foster, L., & Walker, A. (2015). Active and successful ageing: A European policy perspective. *Gerontologist, 55*(1), 83–90.

Foucault, M. (1982). Afterward: The subject and power. In H. L. Dreyfus & P. Rainbow (Eds.), *Michel Foucault: Beyond structuralism and hermeneutics* (pp. 208–226). Brighton: Harvester Press.

Freire, P. (1972). *Pedagogy of the oppressed*. New York: Continuum.

Gladdish, L. (2010). *Learning, participation and choice: A guide for facilitating older learners*. Leicester: NIACE.

Golding, B. (2015). *The Men's shed movement: The company of men*. Champaign, IL: Common Ground Publishing.

Guðjónsdóttir, A. M., Guðlaugsdóttir, I. R., Guðmundsson, H. K., Bru Ronda, C., Aleson-Carbonell, M., & Stanowska, M. (Eds.). (2016). *Towards a dynamic third age: Guidelines and recommendations of the Erasmus + BALL Project*. Poland: Towarzystwo Wolnej Wszechnicy Polskiej Oddzial w Lublinie. https://www.ball-project.eu/sites/default/files/BALL_PublKonc_Internet/Ball_en2_023_200.pdf. Accessed May 25, 2018.

International Association of Universities of the Third Age. (n.d.). *International Association of Universities of the Third Age*. https://www.aiu3a.org/research.html. Accessed February 12, 2019.

Joseph, D., & Southcott, J. (2015). Singing and companionship in the Hawthorn University of the Third-Age Choir, Australia. *International Journal of Lifelong Education, 34*(3), 334–347.

Karpf, A. (2014). *How to age*. Croydon: Macmillan.

Katz, S. (2000). Busy bodies: Activity, aging, and the management of everyday life. *Journal of Aging Studies, 14*(2), 135–152.

Kidahashi, M., & Manheimer, R. J. (2009). Getting in ready for the working-in-retirement generation: How should LLIs respond? *The LLI Review, 4*, 1–8.

Laslett, P. (1989). *A fresh map of life: The emergence of the third age*. London: Macmillan Press.

Midwinter, E. (2007). *U3Aology: The thinking behind the U3A in the UK*. Frank Glendenning Memorial Lecture, 23 March 2007. Leicester: NIACE.

Moody, H. R. (1990). Education and the life cycle: A philosophy of aging. In R. H. Sherron & B. Lumsden (Eds.), *Introduction to educational gerontology* (2nd ed., pp. 23–39). Washington, DC: Hemisphere.

Moody, H. Y. (2004). Structure and agency in late-life learning. In E. Tulle (Ed.), *Old age and agency* (pp. 30–43). New York: Nova Science Publishers.

Moody, H. R., & Sasser, J. R. (2014). *Aging: Concepts and controversies* (8th ed.). Los Angeles, CA: Sage.

Morris, D. (1984). Universities of the Third Age. *Adult Education, 57*(2), 135–139.

Ordonez, T. N., Lima-Silva, T. B., & Cachioni, M. (2011). Subjective and psychological well-being of students of a University of the Third Age. *Dementia e Neuropsychologia, 5*(3), 216–225.

Parsons, T. (1951). *The social system*. New York: Free Press.

Philibert, M. (1984). Contemplating the Universities of the Third Age. In E. Midwinter (Ed.), *Mutual Aid Universities* (pp. 51–60). Kent: Croom Helm Ltd.

Ramírez Gómez, D. (2016). Critical geragogy and foreign language learning: An exploratory application. *Educational Gerontology, 42*(2), 136–143.

Ratana-Ubol, A., & Richards, C. (2016). Third age learning: Adapting the idea to a Thailand context of lifelong learning. *International Journal of Lifelong Learning, 35*(1), 86–101.

Riley, M. W., & Riley, J., Jr. (1994). Structural lag: Past and future. In M. W. Riley, R. L. Kahn, & A. Foner (Eds.), *Age and structural lag* (pp. 15–36). New York: John Wiley & Sons.

Robbins-Ruszkowski, J. (2017). Aspiring to activity: Universities of the Third Age, gardening, and other forms of living in postsocialist Poland. In S. Lamb (Ed.), *Successful ageing as a contemporary obsession: Global perspectives* (pp. 112–125). New Brunswick, NJ: Rutgers.

Russell, H. (2008). Later life: A time to learn. *Educational Gerontology, 34*(3), 206–224.

Third Age Trust (2018). *Learning not lonely. Living life. Expanding horizons. Challenging conventions*. London: The Third Age Trust. https://indd.adobe.com/view/c99ad603-0622-4636-b12c-9f8ad363dae8. Accessed October 17, 2018.

Weber, M. (1978). *Economy and society: An outline of interpretive sociology*. Berkeley: University of California Press.

Wilińska, M. (2012). Is there a place for an ageing subject? Stories of ageing at the University of the Third Age in Poland. *Sociology, 46*(2), 290–305.

Withnall, A. (2010). *Improving learning in later life*. London: Routledge.

Withnall, A. (2012). Lifelong or longlife? Learning in the later years. In D. S. Aspin, J. Chapman, K. Evans, & R. Bagnall (Eds.), *Second international handbook of lifelong learning* (pp. 649–664). Dordrecht: Springer.

World Health Organization. (2002). *Active ageing: A policy framework*. Geneva: World Health Organization.

Wright Mills, C. (1959). *The sociological imagination*. Oxford: Oxford University.

Marvin Formosa is Associate Professor of Gerontology at the University of Malta where he is Head of the Department of Gerontology and Dementia Studies, Faculty for Social Wellbeing, and contributes to teaching on active ageing, transformative ageing policy, and educational gerontology. He holds the posts of Chairperson of the National Commission for Active Ageing (Malta), Rector's Delegate for the University of the Third Age (Malta), and Director of the International Institute on Ageing, United Nations, Malta (INIA). He directed a number of international training programmes in gerontology, geriatrics and dementia care in the Philippines, China, India, Turkey, Malaysia, Belarus, Kenya, Argentina, Azerbaijan and the Russian Federation. He has published extensively across a range of interests, most notably on active ageing, critical gerontology, Universities of the Third Age and older adult learning. Recent publications included *International perspectives on older adult education* (with Brian Findsen, 2016), *Population ageing in Turkey* (with Yeşim Gökçe Kutsal, 2017), and *Active and healthy ageing: Gerontological and geriatric inquiries* (2018). He holds the posts of Editor-in-Chief of the International Journal on Ageing in Developing Countries, Country Team Leader (Malta) of the Survey of Health, Ageing, and Retirement in Europe (SHARE), and President of the Maltese Association of Gerontology and Geriatrics.

Index

© Springer Nature Switzerland AG 2019
M. Formosa (ed.), *The University of the Third Age and Active Ageing*, International
Perspectives on Aging 23, https://doi.org/10.1007/978-3-030-21515-6

Printed by Printforce, the Netherlands